R. C. HICKLING
55 MIDDLEGATE ROAD,
FRAMPTON, BOSTON
LINCS. 0205 722785

WHALING AROUND AUSTRALIA

WHALING AROUND AUSTRALIA

Max Colwell

ANGUS & ROBERTSON (U.K.) LTD
LONDON

First published in Great Britain 1970
by Angus & Robertson (U.K.) Ltd,
52-54 Bartholomew Close, London, E.C.1,
in association with
Rigby Limited, Adelaide, South Australia
Copyright 1969 by Max Colwell
Library of Congress Catalog Number
All rights reserved
Printed in Hong Kong
SBN 207 95246 9

ACKNOWLEDGMENTS

The author gratefully acknowledges the assistance given by the following:

The Bureau of International Whaling Statistics, Sandeford, Norway.

Mr Joh. N. Tønnessen, Oslo, Norway. (Author of The History of Modern Whaling — Norway.)

Royal Norwegian Consulate General, Sydney, N.S.W.

Commonwealth Bureau of Census and Statistics, Canberra, A.C.T.

Department of Primary Industry, Canberra, A.C.T.

Consulate of Japan, North Adelaide, South Australia.

Japan Whaling Association.

Nichiro Gyogyo Kaisha Ltd, Tokyo, Japan.

Taiyo Fishing Co. Ltd, Tokyo, Japan.

Hoko Fishing Co. Ltd, Tokyo, Japan.

The Kyokuyo Hogei Co. Ltd, Tokyo, Japan.

Old Dartmouth Historical Society Whaling Museum, New Bedford, Massachusetts.

Cheynes Beach Whaling Company, Albany, Western Australia.

Nor'-West Whaling Company Pty. Ltd, Fremantle, Western Australia.

Shire Clerk, Shire of Imlay, Eden, N.S.W.

The Public Library of N.S.W.

State Library of South Australia (Reference Library, Research Division, Archives, Newspaper Section).

CONTENTS

ILLUSTRATIONS

*Canst thou draw out leviathan with an hook? Or his tongue with a cord
which thou lettest down?
Canst thou put a hook into his nose? Or bore his jaw through with a thorn?
Shall the companions make a banquet of him? Shall they part him among
the merchants?
Canst thou fill his skin with barbed irons? Or his head with fish spears?*

Job 41 : 1, 2, 6, 7

And Jacob said, Sell me this day thy birthright.

Genesis 25 : 31

1

Every muscle in violent tension

Until the first commercial exploitation of petroleum, in the 1850s, whale oil played an essential part in the economy of the western world. Whales were almost the only source of a pure, clean oil which could be used for illumination, instead of sputtering, smelly tallow candles or the costly beeswax candles. They had been hunted for centuries in the cold northern seas, and as the frontiers of the civilized world were pushed further and further afield, and seamen brought back tales of the great schools of whales in the South Atlantic, Pacific, and Indian Oceans, the whaling industry grew steadily in size. From ports such as Dundee in Scotland and Nantucket in America, the whaling captains sailed their little ships on voyages which sometimes lasted for two or three years, in a long, relentless hunt which did not end until they could turn for home with their holds full of barrelled oil.

It was the hardest possible life. The crews lived in conditions of extreme discomfort, and might not set foot on shore for months on end. When the cry of "*There she blows!*" was roared by an excited lookout, they would row their flimsy boats to grapple at close quarters with the world's largest mammal—far bigger than an elephant, and with the advantage of being in its own element. They had to approach close enough to it for the harpooner to plunge his lance into its side, and the whale's immediate response would be something like that of an upheaval of nature itself. Their boat might be towed for miles by the outraged monster, which might leap out of the water and come down with such a crash that the wave would swamp the boat, or, in the case of sperm whales, it might turn and attack the boat, opening a huge mouth to reveal a gullet which was quite obviously capable of swallowing Jonah.

In most cases, by sheer dogged courage and endurance, the battle was won by the men. Then, they would have to tow the whale back to their ship, and begin the toilsome, odorous process of cutting it up and trying it out, in order to extract the oil. After a few whales had been treated in this way, the entire ship was oily and greasy, thick with the cloying stench of melting blubber, her sails blackened with the thick, sooty smoke from the tryworks fires, which were fed with whatever portions of the whale could not be rendered down into oil.

But it was a profitable business, and for centuries was an important industry—perhaps far more important, by comparison, than it is nowadays, because populations were smaller, whales more frequent (since the great herds had not yet been massacred by the deadlier developments of the twentieth century) and there were fewer competitive products. The industry produced whale oil, for the lamps of those who could afford it and for medicinal purposes; spermaceti candles, and whalebone, whose pliant strength made it esteemed for many uses including the ribs of corsets. These were coveted materials in a world which relied entirely upon the direct bounty of nature.

Consequently, when the First Fleet sailed along the southern coast of Australia, with its cargo of convicts for Botany Bay, the sight of whales blowing their spouts of vapour seemed to promise at least one industry for the infant colony. And, in fact, whaling was to be Australia's first industry, and was in full swing long before anyone had any idea of the nation's vast mineral resources, or had made anything but the most tentative attempts at agricultural and pastoral pursuits.

The great days of deep-sea whaling lasted from towards the end of the eighteenth century to about the middle of the nineteenth. Before then, most whales were caught from small boats putting out from the shore, or by craft which sailed within a few hundred miles of their homeland. Whaling was carried on in places as far apart as Japan and the Faeroe Islands, and the successful kill brought good rewards to the crew who risked their own lives in making it.

Whaling figured largely in the economy of New South Wales, put Hobart Town on the map as a deep sea port, provided South Australia with its first exports, became an escape route for convicts and a cover for smugglers, turned seamen into explorers, and gave Australia its first taste of big business.

As early as 1839, only three years after the proclamation of South Australia as a province, a correspondent of the *South Australian Register* visited the whaling station at Encounter Bay. Over the initials G.B.W., he described what he saw in words which give a vivid picture of the whalers and their methods:

> . . . I was down at the fishery one fine bright morning and talking to Clarke, the headsman, when the whiff or small flag at Rosetta Head was hoisted. The men were lounging about smoking, some in their bunks half asleep, others enjoying a dog-fight or game of quoits, when a shout was heard from the blacks' camp. The flag was only going up when they first saw it, and at their shouting the whole fishery was instantly in commotion. Each man knew his boat and particular duty. The harpoons, lances, lines, and other paraphernalia were always kept ready in the boats, and also a keg of fresh water. The headsman was down among the first, and then the cook rushed to them with the scran-bags. These were waterproof canvas bags, one for each boat, containing beef, pork, damper, and

a good supply of each for they could not know when they might return. All is ready: the extra hands start the boats on rollers, for there was a fixed way or slide for each, and then, with a rush, they are off—three good boats—and to each a crew, well up to their work, and straining every muscle to be before their comrades. The headsman chooses his own crew; and as the same men, if good, are chosen year after year, he generally has the best for a long and strong pull. The whales, a lot of them, have been sighted coming along the coast from the north, and, therefore, out of sight of the boats; but the lookout directs them which way to pull, by holding out a small flag. We scamper along the neck of land at our best paces and climb the steep hill at the end, up a path meandering among huge granite boulders. We gain the top just as the boats come sweeping past and the day is so fine and calm that we can hear the shouts of the steersmen as they goad on the pulling hands to increased exertion. That is unnecessary, for every muscle is in violent tension and the boats seem to spring from the water at every stroke.

The steersman guides with his long oar and sweeps the boat's head wherever necessary and when required, within a very small circle. We had only gained the summit of the Head when the look-out pointed to the north, and, about two miles distant, we saw the blow or spout and could distinguish the huge bodies of first one, then two, in all five fine fish. They were coming leisurely down the coast, right in a line with their enemies, and remained rolling about, occasionally leaping out of the water, and then diving, showing their flukes, and making a report like the discharge of a big gun as they struck the water.

When nearing the Head, and about a mile from the boats, they seemed to change their minds, and, after a longer dive under water than usual, came up with their spouting in a direction right out to sea. This change was at once observed by those directing the boats, which were swept round, and a hard pull in chase was kept up manfully, Clarke's boat keeping the lead, and the other two separated at distances of about one hundred yards all in a line, and evidently the men were straining every nerve. 'There she blows,' cried the look-out, as the jet of water was thrown up like a fountain, 'they can't miss them'; and now commenced such a state of excitement as is seldom seen. He stamped and jumped about, and from his mouth, which contained a great part of a fig of tobacco, there issued such a torrent of strange and fearful oaths that we thought he must have gone mad.

He had a large spyglass, and from time to time gave us information as to how they were getting on. At last we heard that one of the boats was fast to a fish, and then his language and excitement reached the climax, and throwing his sou'-wester on the ground, he danced upon it and ground it into the earth; then taking it up he gave it a slap against his thigh, put it on his head, shut up his glass and commenced filling his pipe, for the boats were all out of sight. It was now noon and we remained smoking and yarning with the look-out, whose blood was up, about whales. In the course of this he gave us some of his experiences and hair-breadth escapes. . . . From the hill we could now and then see specks on the water which were the returning boats, to us almost out of vision, but the blacks could see them distinctly, and the huge monster they were slowly towing to the station. We watched the two which had last left till their long steady sweep had made them also indistinct and then down again to the huts. Here all was bustle. The carpenters, or more correctly, boat-builders, with the coopers and extra hands, were making all ready for the trying out or boiling the oil. The 'tryworks' was a large black shed open on all sides. Down the centre were four enormous boilers; the oil, set in the first, which was over the furnace, was overflowing; it would run into the second and settle, then that full, into the third, and so on. From the last, which was ladled out into the casks.

Just under the 'look-out' there was a large barge, with a couple of derricks rigged, and all the gear necessary for securing and operating on the whale when brought alongside. Everything was in perfect order, and lighting their pipes, the extras strolled away to the Head to see how the boats were getting on with their towage.

They were now well in sight, and the whale with a small flag stuck into it, loomed large. All the boats were in a line, and pulling a long steady stroke, came towards us at the rate of about two miles an hour. It would be sundown before they reached the bay, but we decided to wait for them and get a wrinkle about cutting the blubber from the fish and other delicate operations. The wind, which was from the west, was in their favour, and after a time we heard the song, in which all joined, as they cut their weary way through the sea. About six o'clock they were close into the Head, greeted with cheers by the whites and most diabolical screams from the excited, raging, raving, and jumping natives. The work is nearly over. The boat goes to meet them, carrying a kedge and stout hawser; the kedge is thrown over about three hundred feet from the barge, and the great carcass is pulled alongside and made fast by a toggle through the jaw to the hawser, the flukes or tail is secured by another hawser . . . and then all is bowsed tight. The rolling tackle is passed round the fish and all is secure.

The fish is on its back, and the white belly shows an enormous surface above the water. It puts you in mind of a gigantic pig scalded and all the bristles off. All is now snug and the men go to the headsman's house to get their grog.

And yet the beginning of whaling around Australia was rather slow and somewhat hampered by red tape.

When Captain Ebor Bunker's ship, the *Albion*, was chartered to transport convicts to Tasmania under Lieutenant Bowen in 1803, Captain Bunker had a clause inserted in his agreement which allowed him to put off after whales if they were sighted. Governor King, however, aware of his responsibilities to the Crown and the convicts in his charge, also had a clause inserted: "If boats are put off to catch fish, every convict must be handcuffed and confined below until the boats return."

It is interesting to note that the Governor's concern for the convicts did not prevent him from later privately investing in whaling enterprises, and no doubt he congratulated Captain Bunker on his bag of three sperm whales on the voyage. Whether the good captain chained his convicts below, or gave them the first sight of boiling blubber and a smoking trypot behind which many of their kind were to hide in future years is not known. But neither the presence of convicts nor any amount of red tape could prevent the exploitation of the geographic legacy which began after that voyage.

For centuries before the coming of chained and unchained white men, several species of whales had visited the southern and eastern coastline of Australia; of these the best known to Australian whalers were to be the humpback, right, and the sperm, and later in Antarctic waters, the blue whale.

Much has been written about the size of whales compared with the men who hunted them from frail open boats, but nothing more appealing than the following from Herman Melville's *Moby Dick*: "The fact is, that among his hunters at least, the whale would by all hands be considered a noble dish, were there not so much of him; but when you come to sit down before a meat pie one hundred feet long, it takes away your appetite."

Melville's meat pie was no doubt a blue whale, the bulkiest and largest of all creatures.

Measuring up to 100 feet and weighing as much as 100 tons (old whalers always reckoned on a ton per foot) they have a 1,000 pound heart, a ton of liver, ovaries like soccer balls and pituitary glands (the size of a pea in man) as big as a human fist. Of the reproductive organs, it is sufficient to say they are proportionate to the weight and length of the creature.

But despite their size they swim with ease at fifteen knots and when alarmed at twenty knots, propelling their rather streamlined bodies with alternate up-and-down beats of their horizontal tail flukes. The blue whales' speed made hunting from open boats difficult. Until the advent of steam chasers they were not exploited to any extent.

Compared with the blue whale, the sperm is a slower, grotesque, square-headed monster with a jaw taking up a third of its body. Perhaps because of its ugliness many old whalers had a soft spot for the sperm, and a healthy respect for its fighting ability. It is the only toothed variety in existence, and many a whaling captain used a set of four sperm teeth to keep his charts from rolling back while he plotted a course for their destruction. Weighing about a pound each, they come in sets of between twenty to thirty on each side of the lower jaw. When "Physeter" as he was nicknamed, closes his mouth each tooth fits into a hollow in the upper jaw. Thus he is well equipped to hunt squid, his staple diet, caught by diving into the depths of the ocean.

From the contents of their stomachs, however, it is apparent that sperms do not always make the best use of their teeth as far as chewing is concerned. Whole squids up to thirty-four feet in length, including tentacles, and weighing 400 pounds, have been found before the process of digestion set in. The comparatively large throat and swallowing capacity of sperms no doubt gave rise to the Assyrian fables of men being swallowed by sea monsters, and, perhaps, even to the story of Jonah. Whether Jonah was actually ". . . in the belly of the fish three days and three nights" or whether, as some writers have suggested, he invented the story to explain his absence, will never be known, but one thing is certain; the sperm frequented the Mediterranean in Biblical times and could have been on hand to do the Lord's bidding. They also suffer from intense indigestive pains and are not averse to relieving themselves of unwanted air. On a still day in calm seas, their monumental "burps" can be heard for hundreds of yards. Perhaps Jonah was mistaken for a squid but proved to be a gastronomical irritant, so that when "the Lord spake unto the fish it vomited out Jonah upon the dry land."

Whenever the sperm is mentioned, Herman Melville's albino Moby Dick (a sperm whale) and the sinking of the *Pequod* enjoy a controversial airing. No one can deny that the sperm has the equipment to stove in a wooden ship. Within his square head there can be up to three tons of spermaceti, a colourless liquid which old whalers collected in bucketsful as the best oil available. This for'rard weight, encased in bone, made an ideal battering ram, and when pierced by a harpoon the sperm displayed his irritable nature and armed at both ends, turned on his attackers and either tried to gnaw the boats in two with his teeth, ram them with his head, or smash them to pulp with his tail. In fairness to Melville, mention must also be made of the fact that white sperm whales have been caught up to fifty-five tons and a length of fifty-five feet.

But the best evidence in support of the sinking of the *Pequod* comes from the loss of the *Essex*, a Nantucket whaler under the command of George Pollard, in similar circumstances. On 20 November 1819, when she was in latitude 40 degrees south and longitude 119 degrees west, whales were seen from the masthead. Three boats were lowered, leaving the usual skeleton crew aboard. They went about the usual business of preparing the tryworks, and took little notice when the look-out again saw a sperm whale breaching near the ship; the tryworks were their business, and whales could safely be left to the boat crews. However, when the sperm set a collision course for the *Essex*, the officer in charge ordered hard over helm, but the order came too late. Square-head smashed into the *Essex*, shaking her from stem to stern.

There was nothing unusual in a ship being accidentally grazed by a maddened sperm whale, but after one encounter the creature usually turned tail and beat a hasty retreat, leaving the ship a little the worse for wear. But as Captain Pollard stood in the bows of his whaleboat he saw the whale turn and charge the *Essex* a second time. This time she was hit in the bows and within a few minutes was on her beam ends.

Captain Pollard came aboard, ordered his men to throw gear and provisions into the boats, then pulled away. They lay by all night and went aboard again next morning for additional supplies, but from an inspection they knew the *Essex* was doomed. Captain Pollard, more of a realist than Captain Ahab, did not lay a course for the destruction of the whale which sank his ship, but instead set sail for the coast of South America. During this nightmare voyage they cut flesh from the bodies of their companions who died, before committing them to the mercies of the deep, and when the pangs of hunger became unbearable, drew lots to decide who should be killed and eaten. On 22 February 1820, the Nantucket ship *Dauphin* rescued Captain Pollard and his cabin boy, the only survivors of the captain's boat.

There have been other reports of ships being rammed and sunk by sperm whales, some exaggerated, others unable to stand investigation, but the above was recorded in Captain Pollard's log, and is surely one of the most amazing events in the long history of deep-sea whaling.

Such stories were, of course, relished in every whaler's tavern from Nantucket to Hobart Town, and while Australian bay whalers knew the sperm from reputation and occasionally from actual encounters, they were more concerned with Balaena, the black whale. (The name balaena is the Latin word for "whale.") This species was smaller than the sperm and the blue whale, and when fully grown measured about sixty feet. A docile, sleepy creature, it had the reputation of loyalty and would often sacrifice its own life rather than leave a dying calf. They also came close inshore, and for these reasons were the "right" whales to attack, a name which has stayed with them ever since.

Another species known to Australian whalers, the humpback, which still visits our waters, is classed as a middle sized whale which grows to a size of fifty-five feet. It received its name from the massive thick-set body which gave the appearance of a hump in the water.

Humpbacks, right, and blue whales are known as whalebone whales, and instead of having teeth like the sperm, are equipped with a sieve made up of hundreds of hairy whalebone strips hanging down from the upper jaw. They feed on pelagic shrimp, or "krill," which they take in with huge quantities of water by swimming along with their mouths open. When the water is forced out through the sieve by the closing of the lower jaw, the krill is caught by the plates and is then forced down into the stomach of the whale. It is a case of the largest creatures in the ocean feeding on one of the smallest forms of sea life. The whalebone taken from the mouths of the whales, known as baleen, was used to trim feminine figures and was much in demand by corset and umbrella manufacturers before the invention of mild steel. It was collected as a fringe benefit of the industry by bay whalers all over Australia. In the very early days it was also fashioned into knife handles, and shredded into plumes for the helmets of English knights.

Today whalebone has little value except, it would seem, to water diviners. The Australian *Fisheries Newsletter* of July 1967, reported that: "Recently Mrs B. A. Dalby, a member of the English Dowsing Society, visited Australia and asked the Fisheries Branch of the Department of Primary Industry to help her in the search for suitable whalebone. She explained that whalebone water divining rods are firm and flexible and do not drag and break as hazel twigs do. Tiny flat whalebones sold in some shops for stiffening women's garments are too flabby for the Society's purposes."

During the southern summer, from November to April, whalebone whales stay in the vast food store-house of the Antarctic, but in April with the onset of winter, they head north for the warmer waters along the southern Australian coastline. This voyage is surely the longest sea honeymoon on record; from June to October. During this time they take no food, and are content to go about the business of mating and giving birth to their young. There are several theories about their mating habits, but it appears that they are seasonally monogamous, and are more often than not found in pairs.

The sperm, on the other hand, displays the virility of the meat eater, and often a lone bull will reign over a small harem, with one or two younger bulls lurking in the distance; awaiting the opportunity provided by a well-placed harpoon or natural decline to claim his rightful inheritance.

Apart from lone old bulls, evidently unable to withstand the rigours of harem rule, who find peace and consolation in the Antarctic wastes, sperms prefer the warm water of the sub-tropics and tropics. But they, too, follow a set path of migration, in semi-clockwise sweeps across the Pacific, which brings them to the coast of Australia.

As well as instinctive migration, which could be relied upon to bring the whales within reach of Australian whalers, another habit was of vital interest to those who relied upon them for a living. All whales breathe through a blowhole situated on top of the snout. In the quotation at the beginning of the chapter, the blowing of the whale is described as "the jet of water thrown up like a fountain." Whales, however, do not blow water, as there is no connection between the lungs and the digestive passages. What comes out is a warm blast of the foulest smelling air imaginable, which expands rapidly when expelled through the blowhole. This expansion causes a lowering of temperature, leading to condensation of water vapour in the expelled air. This is what old whalers thought was a spout of water.

These, then, were the creatures which were the raw material of the Australian whaling industry; warm-blooded mammals with lungs and four-chambered hearts, which fed their young with milk. The early whalers simply called them "fish," and they came year after year. Wherever their spouts were sighted either from a lonely bay whaling station or a masthead look-out, the chase which followed certainly put "every muscle in violent tension."

2

I got clear of my live lumber

GOVERNOR ARTHUR PHILLIP, THE COMMANDER OF THE FIRST batch of unwilling colonists to be landed in New South Wales, became deeply, though at times unwillingly, involved in the foundation of the Australian whaling industry. When the First Fleet dropped anchor off Sydney Cove, whales were a common sight off the Australian coast, and it is possible that they quite often strayed into Sydney Harbour. At least one of them did so shortly after the establishment of the convict settlement, and led indirectly to the incident in which Governor Phillip was speared by an Aborigine. This happened in September 1790, and the following eye-witness account was written by James O'Hara:

A general attention was excited at this time by the entrance of a spermaceti whale into the main harbour. Those who were best acquainted with the nature of this enormous animal, were not without a degree of alarm, which afterwards proved to be but too well founded. He differs, it seems, much, from his dull kindred of the north seas, and in nothing more than in a disposition to mischief. Several boats were gallantly put off with harpoons, but

13

from the unskilfulness of the adventurers returned with the sole success of having escaped him. He continued to range through the harbour, and was shortly the cause of a very melancholy event. Three soldiers, who had been for some days fishing, called at the look-out, where they took into their boat Mr Ferguson, a midshipman. The whale, whether in his gigantic play or from an instinct to destroy, pursued the boat and overset it. One of the soldiers reached the beach by swimming; the three other persons lost their lives by this strange casualty.

This dangerous inmate continued to infest the harbour for several weeks. At length, however, he ran himself ashore in Manly Cove, where he soon attracted whole tribes of natives, who flocked there from all parts to regale upon him.

The governor was at this time on a fishing party on some of the shores in that vicinity. It is remarkable that the day before the running aground of the whale, there were two of the most extraordinary hauls of fish ever known in the colony. In these two hauls were taken nearly 4,000 of a species, called by the colonists, salmon. The runaway native, Bennillong, appearing at some distance, but being fearful of approaching, had 30 or 40 of them sent to him as a sort of assurance that he was forgiven, and as an overture to his return. The governor was then induced by curiosity to go to the whale feast, where Bennillong also came, having so far laid aside his fears as to speak to him, and to act upon the occasion as a kind of conductor of the ceremonies of the place, introducing successively, by name, and with an importance truly ludicrous, a number of the more distinguished natives assembled there.

The scene, however, soon changed to a very serious one. A native, who had hitherto stood at the distance of twenty or thirty yards, having been pointed out, the governor advanced towards him amicably extending his hand; upon which the savage withdrew with an air of distrust, and the governor following him with a continuation of the same action, he took up a lance from the grass and darting it at him with force and exactness, it entered a little above the collarbone and pierced through. It was barbed. The arms were in the boat—the cockswain leapt out and fired a shot, which however did no execution. He was followed by three others whose arms snapped and missed fire. The savages dispersed. The governor now retired with the gentlemen that were along with him, to the boat. They had some miles to row before they reached the settlement, but the men exerting themselves to the utmost, he was at his house in two hours.

The weapon was extracted by Mr Balmain, assistant-surgeon of the hospital, who pronounced the wound not mortal. His sufferings did not prevent the governor's issuing the most express orders against any attempt at retaliation—thus might this same spermaceti whale (so oddly linked is the chain of human events), have been the not very remote cause of the death of the governor, after having been the immediate one of that of three other persons.

It is apparent that, although the whalers mentioned were inexperienced, whaling equipment was available in the colony. That first sperm whale may have had the run of the harbour but others which put in an appearance in the later years certainly did not. But before this could happen, outdated laws and monopolies had to crumble.

At the time of Phillip's arrival, the East India Company held a monopoly of trade from the Cape of Good Hope to the Straits of Magellan. The Governor had been instructed to forbid any contact between his colony and any other country. This was understandable for a penal colony, but it tied his hands as far as trade was concerned.

But captains of East Indiamen had seen large schools of sperm whales in the Indian Ocean. As merchantmen they could afford to ignore them, so too could the North American whalers from Nantucket, and British whalers who were doing quite well in the North Atlantic. But when the American War of Independence began, the British ships were forced further south into the South Atlantic. Here they met moderate success, but they believed that warmer waters would yield greater catches. In 1788, the year that the first fleet dropped anchor in Botany Bay, the British whaler *Emilia*, owned by Sam Enderby & Sons, nosed her way around Cape Horn and entered the Pacific. She returned to London with a full cargo of sperm oil, proving that the Indian Ocean grounds extended across the world.

This whetted the appetite of other British whalers, and in 1791 five ships, which had carried convicts to Port Jackson, planned to sail for the coast of Chile and Peru after discharging their human cargo. One of them, the Enderby ship *Britania*, was under the command of Thomas Melville. On 22 November 1791, while his ship lay at anchor in the harbour, he wrote to her owners in London: "I have the pleasure to inform you of our safe arrival in Port Jackson in New South Wales, October 13th, after a passage of 55 days from the Cape of Good Hope.... After we doubled the southwest cape of Van Diemen's Land we saw a large sperm whale off Maria's Islands but did not see any more being very thick weather and blowing hard till within 15 leagues of the latitude of Port Jackson. Within 3 leagues of the shore we saw sperm whales in great plenty. We sailed through different shoals of them from 12 o'clock in the day till sunset, all around the horizon, as far as I could see from the mast head . . . In fact I saw very great prospects in making our fishery upon, this coast and establishing a fishery here. Our people was in the highest spirits at so great a sight and I was determined, as soon as I got in and got clear of my live lumber, to make all possible despatch on the fishery on this coast."

The excitement which Melville felt at sighting the whales made him repeat himself in his letter, but he was as good as his word and "waited upon His Excellency Governor Phillip."

The Governor had decided to despatch the *Britania* and her convicts to Norfolk Island, but Melville, "immediately told him secret of seeing the whales thinking that would get me off going to Norfolk Island." The Governor must have been on the horns of a dilemma. His colony needed exports and industries, the convict ship captains were chafing at the bit of the East India Company monopoly, and yet he knew that the seas east of the Cape were out of bounds to them. Melville, however, pressed his case and must have been quite persuasive. Phillip reversed his decision on Norfolk Island, and, as Melville points out, "he took me into a private room, he told me he had read my letters and that he would render me every service that lay in his power; that next morning he would despatch every long boat in the fleet to take our convicts out and take on our stores immediately, which he did accordingly and he did everything to despatch us on the fishery."

But there were other ships in the harbour, and Melville lost the advantage. "The secret of seeing the whales our sailors could not keep from the rest of the whalers here, the news put them all to the stir, but have the pleasure to say we was the first ship ready for sea. We went out in company with the *Will and Ann*, the 11th day after our arrival."

Governor Phillip, in writing to Secretary Stephens on 16 November 1791, said: "The great number of spermaceti whales seen on this coast give reason to hope that a fishery may be established here, and several of those ships intended for the north-west part of America are gone to the southward in search of fish, the master of the *Britania* having declared that he saw more spermaceti whales between the south cape and this harbour than he saw in six years on the Brazil coast."

And so the Governor made the decision which opened the way for Australia's first exporting industry. The *Britania* was the first of five convict ships which got rid of their live lumber to go a-whaling; the others being the *Mary Ann*, *Matilda*, *William and Ann*, and the *Salamander*.

The weather, however, was not kind to our first whaling fleet, and Governor Phillip must have been disappointed when he wrote to Under Secretary Nepean on 18 November 1791: "The *Britania* has returned after having been out for 15 days. The master says he saw a great number of fish, and had many in sight for 9 days, during which time the badness of the weather prevented his putting a boat into the water. The day after he left the harbour, in company with the *William and Ann*, seven fish were killed . . . but a gale of wind then coming on, only one fish was saved by each."

17

It was unfortunate that Melville encountered a storm; the whales were there in plenty, the colony offered port facilities, and if a fishery could have been established the long voyage to South America would have been unnecessary. But sperm whaling was a calm weather operation, and although a few convict ships tried their luck in local waters, most left for Peru after landing their convicts.

On 11 October 1792, Governor Phillip expressed his opinion on the matter: "I believe not one of them gave the coast a fair trial, nor can I suppose that they left it solely on account of bad weather and strong currents. The weather on the coast of Brazil is not better than it is on this coast, nor have the whalers there those advantages of harbours which ships employed on this fishery would have; as to the currents, they are pretty much the same on both coasts."

But if rough weather threatened to rob the new colony of an industry, another storm was to play into its hands. War broke out with Spain, and when Spanish cruisers put in an appearance off Cape Horn the whalers returned to the Australian coast. So, too, did the American whalers. From the early days of the colony they had arrived with cargoes, hoping to dispose of them, and as early as 1802, the American brig *Fanny* passed through Bass Strait. Another whaler, the *General Boyd* under Captain Bunker, also visited Port Jackson in that year.

From a broad reading of Australian history, one usually gains the impression that British ships and captains were always to the fore when it came to landing on Australian soil after the arrival of the first fleet, but the Americans were never very far behind. The *Sydney Gazette* of 15 May 1803 reported the arrival of the ship *Good Intent*, whose captain, J. Chase, said that he had met an American, Captain Pemberton from New York, in the brig *Union*. Pemberton had fallen in with the French Captain Baudin, who met Captain Flinders at Encounter Bay on the South Australian coast on 8 April 1802. While in southern waters, Pemberton landed on Kangaroo Island at the foot of St Vincent Gulf, and from local timber built a small schooner which he named the *Independence*. This craft, built in South Australia thirty-four years before the arrival of the first settlers in the southern colony, was eventually sailed to Sydney.

From that time on, the American invasion increased rapidly, and by 1840 Commander J. Stokes in his book *Discoveries in Australia* reported that thirteen American whalers were lying at anchor at Swan River, Western Australia.

The French were also on the scene. On 2 June 1841, the explorer Edward John Eyre and his native companion Wylie were trudging around the coastline of the Great Australian Bight, attempting to find a stock route between Adelaide and Perth. Exhausted and famished, with the rest of their party having been killed by treacherous natives, they looked out to sea and saw, of all things, two whaleboats of the French whaler *Mississippi*. Under the command of an Englishman, Captain Rossitter, they had set up a shore station near what is now called Esperance Bay. For ten days, Eyre and Wylie enjoyed the whalers' hospitality before going on to Albany. If it had not been for Eyre's journals, this little settlement on the western Australian coast would probably have gone unrecorded. Unlike the British, the Americans and French, who also engaged in sealing operations, were not hampered by the East India Company monopoly, and such terms as "territorial waters" were completely foreign to them. They were free to roam and fish at will, to the detriment of the struggling Australian industry.

But if the British whalers had a lukewarm friend in Governor Phillip they were soon to have a champion in Governor King. On 14 August 1804, he wrote to Lord Hobart: "In Lieutenant-Colonel Collins' last letter to me from the Derwent he informed me that the master of an American ship was building a vessel in Kent's Bay on Furneaux's Island, at the east entrance of Bass's Straits. This is the third American vessel that has within the last 12 months been in the Straits, and among the islands procuring seal skins and oil for the China market. In a former letter I requested instructions respecting how I was to act with vessels belonging to powers in amity with His Majesty who resorted to these straits and the islands lying in them . . . I respectfully request being informed how far the Governor of this territory would be justifiable in preventing this intrusion and intercourse with the Americans, which is not only pernicious to the public interest, but highly disadvantageous to the adventuring colonists . . . The enclosed proclamation will show your Lordship that I have taken every step I considered myself justifiable in adopting, and by my orders to Lieutenant-Colonel Paterson it will be observed the precaution I have used that no act of hostility might be urged by the Americans; but as the adventurers of this colony who I have ever made a point to encourage in the fishing and sealing, are not at ease on this point, I humbly suggest the necessity of instructions being sent on this head, as it is almost the only present means they have of benefiting themselves and the colony and in which they have adventured much.''

And so Governor King made his stand; friendliness towards the Americans, but protection for the new industry of the colony. But this was only the beginning. He had also been pressing for the lifting of the East India Company monopoly, using the presence of foreign vessels as a lever. He was not alone in this battle. The firms of Sam Enderby & Sons, and Messrs Champion, were also busy writing letters to the government. They emphasized the benefits of a large number of ships sailing for the colony, the employment of men needed to refit them for whaling voyages, the general boost in trade, and the danger of leaving whaling grounds open to foreign powers. Thus increasing pressure was brought to bear on the powers that be.

Slowly the barriers came down, and by 1820 Australian waters had been cleared to British whalers and merchantmen. Apart from this, Governor King also placed restrictions on the Americans, and in June 1804 he issued general orders which prevented their vessels from using Port Jackson as a base for their whaling vessels. He was not successful, however, and the Americans continued whaling along the Australian coast in increasing numbers. The new industry was to be carried on in the face of keen competition, but this was typical of deep sea whaling all over the world, and it

In the early days, most Australian whaling was shore-based. This old print shows boats tackling a pod of whales in Two Fold Bay (*Mitchell Library*)

The headsman stands ready with the harpoon, while the whaleboat crew pulls towards a surfacing whale (*Mitchell Library*)

BAY WHALING

off the Boyd Town Light House, Twofold Bay, N.S.W.

would have been most unusual if it had been any different on the Australian coast.

The presence of foreigners, this time French, in Australian waters, however, was indirectly responsible for another branch of the whaling industry being established. On 9 May 1803, Governor King again wrote to Lord Hobart: "The *Cumberland*, colonial schooner, which I sent to the southward, returned here on 8th March. By her I received a letter from the Commandant of the French expedition of discoveries, a copy of which—with my remarks thereon—I have the honour to enclose. By its tenor your Lordship will observe that he does not avow having instructions to make any settlement on Van Diemen's Land. What intentions the French Government may in future have on that island I cannot pretend to say further than I have communicated to your Lordship; but I respectfully conceive some instruction should be sent on that head, as it is within the limits of His Majesty's territory."

King only waited four months and in September, 1803, the first settlement under Lieutenant John Bowen was established at Risdon Cove on the River Derwent and in May 1804 another colony was set up in the north on the Tamar.

It was a peaceful spot that the settlers chose on the Derwent; the broad expanse of water, the snow-capped peak of Mt Wellington in the distance, the timbered shore line and the moderate climate must have seemed ideal, for free settlers at least. But again it was an isolated community, faced with the problem of feeding itself in a virgin land. In the normal course of events, cottage industries would have made their appearance as the population increased, but it would be safe to say that the wildest dreams of these first settlers would not have visualized the changes that were to take place on the Derwent.

It is possible that in September 1803, many whales were still spouting in the river, but the settlers could not have seen the full influx of the black whale until July next year. Then they would have appeared at first in ones and twos, cows bulging with calves and bulls lurking in the distance

Ships at anchor in East Boyd, on Two Fold Bay. This was the site of Australia's first "industrial complex," based on the whale fisheries (*Mitchell Library*)

Ben Boyd's yacht *Wanderer*, in which he sailed from England to Australia to found his financial empire (*Mitchell Library*)

awaiting their opportunities. Some would come close in shore, others remain in midstream, a hazard to shipping, but as the days passed their numbers would have increased enormously. And here the settlers would have been agreeably surprised. They were not mischievously-minded sperm whales, of the type which wrecked the boat in Port Jackson, but black whales, sleepy, and inoffensive by comparison with the sperm.

This yearly migration could have only one result. William Collins, a Royal Navy master who came out with the idea of engaging in the whaling industry, and who had been appointed Harbour Master in the new colony, laid out plans in August 1804. He set these before Lieutenant-Colonel David Collins, Governor of the settlement, who read: "The River Derwent is most advantageously situated for the establishment and carrying out of a South Sea whale fishery . . . two or three ships, not exceeding 250 tons each, should be fitted out with every article that has been found necessary for the purpose of a South Sea sperm whale fishery . . . ships fitted out in the above manner should arrive here in the months of September or October, and the ships got ready to proceed to the sperm whale fishing ground so as to be there the first of the season, which commences in December and continues till April. By the time of their return everything might be got ready at the factory to receive their cargoes, and in time sufficient for them to pursue the black whale fishing which commences here early in July and continues till September."

William Collins also added an interesting note about the migration habits of black whales: "I point this out as one of the great advantages that a concern fixed here would derive, as every opportunity might be taken of killing black whales, having frequently seen them out of the regular season in considerable numbers." They could certainly be relied upon to put in an appearance in season, and apparently at odd times as well.

In his plan to combine deep-sea sperm whaling with bay whaling, William Collins hit upon the ideal combination, and one which came into being soon after colonization. In

1806 he established a bay whaling station at Ralph Bay on
the east side of the Derwent, now known as Trywork Point.
Others soon followed his example, and stations were established at Tinder Box Bay, Trumpeter, and Adventure Bays.

Thus the foundations were laid on the Derwent, and in
Port Jackson, for the golden years of Australian whaling.

3

Currency lads partial to the sport

"MOST OF THE BAYS AND HARBOURS ABOUND WITH RIGHT whale . . . and the inhabitants of Hobart Town have frequently witnessed the whole method of killing them from their windows." This was part of A Description of Hobart Town for Emigrants, in Godwin's *Guide to Van Diemen's Land* (1823). While this spectacle must have been repulsive to some, it was certainly encouraging to others. Unlike deep-sea sperm whaling, which required a ship, a trained crew, equipment, and a captain who knew his business, bay whaling, which sprang up on the River Derwent in the early years of settlement, was in reach of the ordinary man.

In 1833, Henry Melville wrote: "Another invaluable article to the colony, as an export, is oil . . . it may be remarked that the situation of the Island and the numerous nooks and bays with which it abounds, render it such a place of resort for whales throughout the winter, that the equipment of a few boats and the erection of a boiler or two upon shore for rendering down the oil, are nearly sufficient, as the outfit of what may be considered with tolerable certainty, a profitable enterprise."

And this is just what happened. Anyone with sufficient capital to buy a whaleboat and equipment was never short of a crew. Labourers, runaway sailors, and farmers who might never have pulled an oar in their lives found that there was more money to be made in whale-oil than from grubbing dirt or swabbing decks.

And so, at the beginning of each season, whaleboats loaded with trypots, harpoons, lances, tubs, and other whaling gear, with provisions for three or four months, set sail from Hobart Town. They could either apply to lease a section of land or, if they believed they were far enough away from authority, simply take over a likely spot. This was usually a sheltered bay or inlet, handy to high ground which could be used as a look-out.

A few rough shelters were then built on shore; one for the cook, another to cover the trypots, and one for perishable stores. If necessary, a timber ramp would be put down into the sea for hauling the blubber up to the tryworks. Then they would post a look-out, or if they became impatient row out into mid-stream to be ready. Only the coopers, the cook, and the carpenter stayed ashore.

The most valuable part of their equipment was the boats; clinker-built from cedar planks, about thirty feet long, pointed at both ends, and higher out of the water at the bow and the stern than amidships. A short post, the loggerhead, projected through about five feet of stern decking. This was used for checking the line as it ran out after a strike. Quite often the line would smoke if the whale dived, and would have to be watered with a bailing dipper kept in the stern. Some headsmen cut notches in the loggerhead for every kill.

The boats had thwarts for from five to eight oars, but usually only pulled five, so that when the harpooner left his oar, to stand in the bows ready to make a strike, the pulling was even on both sides. He could brace himself by resting his thigh in a niche cut into the for'rard decking. The headsman stood in the stern, wielding a twenty-seven foot sweep held in place by a leather strap. He steered the boat, and if necessary could use the sweep to help propel it through the water.

The oars were generally "peaked" if the boat was being towed by a whale, the hafts being placed in sockets on the sides of the boat.

Another important item was the line, of which up to 200 fathoms was meticulously coiled in two tubs so that it could run out free and fast. It came from the tubs aft, around the loggerhead, then for'rard to pass under an iron bar in the bows and back again to be attached to the harpoon. Passing it under the iron bar caused the boat to be towed by the bow instead of sideways during the whale's attempts to escape after being "pricked" by the harpoon.

Boat stores usually consisted of a keg of water, spare lances and harpoons, grog, coats, biscuits, compass, and candle or oil lamp. Thus equipped, the Derwent River bay whalers were ready for the hunt, and by 1841 the smoke from their trypots hung over no less than thirty-five stations in the Derwent and around the Tasmanian coast.

But the authorities did not co-operate with the early bay whalers on the Derwent. In fact, they seemed to do their best to discourage them. A duty was imposed on oil landed in the colony, and came very close to ruining the industry in the early days. It was part of the regulations designed to keep shipping out of the Derwent, and thus make it difficult for convicts to escape. Ships' captains were forced to take their rudders ashore and unbend their sails, and no passengers were permitted to be landed after sunset.

Even the authorities must have known, however, that the "safety" of the convicts would have to take second place to an industry which was spreading across the world. Shipping captains complained, bay whalers complained, and merchants complained. Consequently, in 1813 the restrictions on shipping were lifted, and in 1823 the heavy duty on oil was eased. The shackles had been removed from the industry and in some degree from the convicts, and both made the most of it in the years that followed.

On 24 August 1832, the *Colonist* newspaper listed the owners of the bay whaling establishments, including Mr Hewitt's party in Recherche and Adventure Bays, which took thirty-four fish, and went on to give an account of the season: "The fish are exceedingly plentiful on our coast this season, and have not been known to be more numerous at any former period. It is a most fortunate circumstance, in the present impoverished state of the colony, that the whale fishery is likely to furnish our merchants with an article of export to remit to England, in place of British silver, which has disappeared, or Treasury Bills, which cannot be purchased."

By this time the skill of the bay whalers had improved, so that delighted merchants and writers said that "the currency lads, as the country-born colonists in the facetious nomenclature of the country are called, in contradistinction to those born in the mother country, are extremely partial to the sport, and are bold, active, and expert."

Another reference, this time by Henry Melville in 1833, also tends to liken the business of bay whaling to a sporting pastime: "Thus, almost at our very door or threshold, are we provided with the means of becoming rich, with little comparative trouble or exertion; and at the same moment, are we rearing up a fine and manly race of native youths, in a nursery that would qualify them to contest the palm of superiority on the water, with the inhabitants of any country upon the whole face of the globe."

In actual practice, bay whaling was a backbreaking, dirty, lonely, and often dangerous occupation. They may have been hunting sleepy right whales heavy with calves, but one smash of the tail in a mad flurry could put everyone in the water. And competition was often fierce. An inlet may have been deserted when a party arrived, but as the season progressed, two or even three parties may have decided to settle in it. This meant that five or six boats would close in on one whale.

From a distance there may have been an element of sport in this, but the whalers were in deadly earnest and the stakes were high. Any unwritten laws followed were traditional, with their foundation in the rule that possession was nine-tenths of the law and the term "fast fish," meaning one that had been harpooned, echoed across the bays and inlets if a rival harpooner tried to secure a line to another's fish. But lines often broke and the whale then became a "loose fish" open to attack again from all boats. Lines could also be cut by a second party, and although this was decidedly frowned upon by whalers there was no act to regulate fishing, and no authority to enforce fair play.

These terms "fast fish" and "loose fish" were eventually incorporated in an 1838 act passed in Tasmania, which formed the basis for acts to regulate fishing in South Australia. (See chapter six.) But in the early years in Tasmania, disputes were settled on the beach or out to sea with oars and harpoons.

Tasmanian whalers may have been "a fine and manly race" but bay whaling was certainly not an attractive pastime.

Nor was it, judging by the agreement signed between men and owners, a 40-hour week occupation. One paragraph reads: "And each of the said several seamen hereby promises that he will diligently and faithfully do his duty by day and by night, during the continuance of his term of service under the agreement and obey the lawful commands of the owners."

Ideal conditions were calm weather and good visibility, but moonlight hunts were also practised. So too were forays to cut away harpooned whales marked and secured close in shore for cutting up. Many a crew laboured long in towing a carcass back to shore only to wake next morning and find it in the trypots of a rival party. No wonder feelings ran high.

Perhaps this is why owners had the night work clause inserted. They were more interested in results than unwritten laws, and turned a blind eye to the methods used by their headsmen. But to protect themselves they also inserted another clause in their agreements: "And it is agreed that absence from the fishing station or vessel, boat, or party for more than 12 hours without lawful excuse shall be deemed a desertion." If the fishing was good, deserters were no problem, but if it wasn't then the owners could expect some of their men to move on or turn to other pastimes. These were grog and native women, and it would seem that the latter were quite willing to exchange the life of the bush for the life of a fishing station. In 1823, Godwin's *Guide* included this report on the local Aborigines:

Their skins are black, their hair woolly, and in their features and appearance they resemble the Negro; they mostly go naked, but some of them have a kangaroo skin slung over their shoulders; they have no houses, and lead a wandering life, depending upon hunting and fishing for sustenance, and are perpetually at war with each other. The women are sometimes known to run away from their husbands, owing to the hardness and tyranny they exercise over them . . . and, they say they find their situation greatly improved by so connecting themselves with the whaling gangs, for their native husbands made them carry all their lumber, and perform all kinds of hard work. They have always proved faithful and affectionate to their new husbands, and seem extremely jealous of a rival; the children produced by an intercourse with the natives and Europeans are handsome, of a light copper colour, with rosy cheeks, large black eyes, and well-formed limbs.

But while bay whalers were making the most of their loneliness and small ships were gliding back and forth taking their barrelled oil to Hobart Town, others had turned their attention to the sperm whale. British, American, and French ships had discovered the New Zealand whaling and sealing grounds and began using Hobart Town as a refitting base. Their masts and spars dominated the harbour and on Good Friday 1847 no less than thirty-seven whalers were waiting for a refit.

Local merchants and business men, however, were not to be outdone. They knew that to compete they would have to build their own ships, and there was no better place for it than their own island; huon pines for masts and spars, and blue gums for hulls, grew there in abundance.

The first Tasmanian whaler launched was the *Maria Orr*. Of only 289 tons, she was eventually sailed as far as the Kodiak Islands of Alaska. Many others followed, and by 1849 there were forty-two whalers sailing out of Hobart Town, employing over 1,000 men. Meanwhile, Sydney merchants were also acquiring their own fleets and one, Bobbie Towns, boasted thirteen ships by 1851. Smaller merchants staked their fortunes on one or two vessels.

During this period Australian whaling grounds were divided into four sections; the Middle Ground, between Sydney and New Zealand; the Northern Ground, between Queensland and New Caledonia; the Western Ground, stretching from Tasmania to Cape Leeuwin; and the Eastern Ground, east of Tasmania.

And so bay whaling and deep-sea whaling gave the two colonies the industries they needed, so that a *Complete Guide to the Emigrant*, published in 1832, could say: "To many emigrants, Van Dieman's Land offers many advantages, at least to those who have a small capital. To mechanics generally, even without capital, it affords, with sobriety and industry, a comfortable and certain livelihood, and the means of speedily becoming land owners. Carpenters, builders, smiths, masons, boat-builders, tailors, shoe makers, and indeed all useful trades are in much request . . . boys of ten years of age can earn about three shillings a week and their meat, and older ones in proportion."

Apart from industrial opportunities, the same guide listed other advantages: "In such a climate and with the active life which settlers, in a new colony must necessarily lead, the health of the inhabitants, as might be supposed, is of the best kind. The atmosphere is for the most part dry and elastic, and though it has not as yet, that we know of, been correctly analysed, yet it certainly contains a larger proportion of oxygen than most countries of the world. The stimulating effect of this gas taken into the lungs, naturally communicates itself to the stomach, and tends to keep in a healthy state, the digestive action of that grand organ."

So it would seem that all was healthy in Hobart Town during these formative years. Industry, bracing climate, and opportunity were all there to be enjoyed, but the deep sea whalers who frequented the settlement were a race apart. Wherever they made a temporary home on shore, they left their mark.

They smuggled grog and tobacco ashore to trade with the

locals, and as visitors they had a certain immunity from arrest. It was much easier and cheaper to return a drunken sailor to his ship than to gaol him for a petty offence. The streets took on a cosmopolitan air; American whaling ships roamed far and wide, and very few whalemen returned on the same ship in which they sailed. Whalers picked up crews where they could find them; Portuguese rubbed shoulders with Negroes, Maoris and Scandinavians sat at the same tables with Central Americans.

Not that the locals were noted for their temperance, as the Hobart Town *Courier* pointed out in 1831, "The quantity of spirits and other strong drink consumed annually in the colony may, on a moderate computation, be taken at not less than 100,000 gallons, which, according to the population, allows the enormous quantity of about five gallons to each individual, young and old, male and female in the island." The same writer stated that there were four breweries, and that "the only fault that can be found with the ale or beer, is, that in consequence of the proprietors not being fully able to meet the demand, it is sold and consumed before it is old enough." He also lists some of the more prominent taverns, including the Derwent Hotel, Waterloo Tavern, the Ship, Dallas Arms, and the Commercial.

Rowdy brawls, smuggling, and occasional violent crimes had to be tolerated as the price of prosperity brought by whaling, and were no doubt shrugged at by everyone— except those who had no stake in the industry.

The task of a clergyman in Hobart Town was particularly arduous. In 1832, it was stated in the *Complete Emigrants Guide*, that: "He is placed as it were in the very gorge of sin, in the midst of the general receptacle for the worst characters in the world, and of necessity compelled to take the 'Bull by the Horns,' to grapple at the very gates of hell, if he would rescue a soul from the headlong ruin to which he is hurrying. He has here to struggle with the very enemy at close combat, face to face, and foot to foot, and to brace himself up to the utmost point of exertion."

But the same writer hastened to assure intending immigrants that, "there is probably a larger proportion of intelligent well-informed persons in Hobart Town than in most communities of the same size in any part of the world. This arises in part from it being the seat of Government, and from the number of respectable persons whom it necessarily employs and draws around it, from it being the capital of the island, and the grand focus, as it were, of consumption on the one side and of supply on the other, with all parts of the interior, thereby affording profitable employment for a considerable body of commercial men, and from the circumstance that few but men of intelligence and ability, of strong minds, and bold spiritual enterprise would venture in the first instance to undertake the long and arduous voyage to so remote a country, and to struggle with the numerous and unavoidable difficulties, the hardships and privations incidental to a new colony."

But there were also those in and around Hobart Town who did not undertake to make the long and arduous voyage of their own free will. To the convicts, whether chained or on parole, the whaling industry also had its attractions.

How often they must have stood and gazed at the masts of ships at anchor in the estuary, and watched as their crews rowed ashore to commit crimes which were often worse than those for which the convicts had been transported. They knew that every ship that dropped anchor in Hobart Town would lose one or two of its crew by desertion. Deserters had to be replaced, and fo'c'sles of foreign whalers were always open to any one who could climb aboard, with no questions asked.

They oiled their wrists, in an attempt to slide them out of their manacles, or fought over small pieces of metal which they could hide on their bodies and use later to loosen their irons. Once free, they ran off into the scrub, and when the search died down slipped into the cold water and made for the whaling ships. And in the deserted bays headsmen occasionally woke to find one of their boats and a quantity of stores missing. When the first boat called to pick up their oil it would be reported to the authorities. But where could they look? Whale boats were seaworthy, and in the hands of desperate men could travel a long way in a few days.

These runaway convicts became the first unofficial explorers of much of the southern coast of Australia. Many settled on the islands of Bass Strait, where they eked out a living by growing a few vegetables and trading seal skins for food and grog with foreign whalers.

Those who escaped on foreign vessels were taken further afield, and in 1836, when the first official settlers landed on Kangaroo Island at the foot of St Vincent Gulf, a man who called himself Governor Wally was found living in a hut surrounded by a well-kept vegetable garden.

G. French Angas, in his book *Savage Life and Scenes* (1847), speaks of a man who had been on the island for twenty-one years. This means that he must have arrived in 1815 or 1816. When Captain John Hart visited the island in 1831, as the master of the schooner *Elizabeth*, he found eighteen men, with native wives who had been "recruited" from the mainland. Since they were equipped with whaleboats, it is quike likely that the early inhabitants of Kangaroo Island had visited the mouth of the River Murray, some thirty-six hours sailing time from their home, long before Captain Charles Sturt saw it in 1829, after his gruelling thirty-three-day journey downstream in a similar boat.

And it is also likely that Captain Collet Barker, explorer of the Murray mouth, was speared by the Aborigines in revenge for the Kangaroo Island men's treatment of native women. They killed him as he swam across the Murray mouth on 30 April 1831.

From the Derwent, bay whaling also spread to the western parts of Victoria. As early as 1828 whalers and sealers had built their shacks and gardens near what is now called Portland Bay. But like other isolated settlers they had no legal right to the land and were not recognized by the British Government except as intruders. But once again when organized settlers arrived (in the case of Portland Bay, the Henty Brothers) they found some form of civilization to greet them. Bay whaling continued along the Victorian coast out from Portland in the same way as it did in the Derwent and by 1840 when more land had been opened up, it was legalized by the British Government.

And so bay and deep sea whaling made its mark on the Australian scene. Hobart Town became the focal point of the industry from where adventurers set out to more distant grounds. As W. J. Dakin pointed out in his book *Whalemen Adventurers*, it was so much part of life that newspapers often included whaling news with more exciting material and he quotes a report in the *Hobart Town Gazette* of 29 June 1816:

"A great number of whales have already made their appearance in Frederick Henry Bay; some few have been seen as high as Sullivan's Cove—Preparations are making by Mr D. McCarthy and Coadjutors to begin the fishing— the different elements contribute to our prosperity, when industry leads the way. On Thursday last by special license, was married by the Rev. R. Knopwood at the Derwent Hotel, T. W. Stocker to Mary Hayes, widow of the Derwent Hotel, Elizabeth Street, after a tedious courtship of two years."

The newspaper gives no clue as to why the courtship was tedious, and one can only assume that hotel widows were sometimes as difficult to harpoon as right whales. But it does make one point; whales were still presenting themselves twelve years after the industry began and during that time it had bred some remarkable men.

Captain James Kelly was one. Tall, rugged, and forceful, he was born in Sydney in 1791, the year that Thomas Melville waited on Governor Phillip for permission to go sperm whaling. He was a true son of the whaling industry, and before he was twenty-eight had rounded Tasmania in a five-oared whaleboat and explored Port Davey and Macquarie Harbour. At the height of his career in 1819 he was made Harbour Master and Pilot of Hobart Town, at a time when keen rivalry existed in the industry. Previous to this, he and a man named Hewitt had sent a party to Portland before the arrival of the Hentys.

But Kelly was a business man, and knew that heavy port charges and duties on whale-oil were placing severe strain on the industry. In 1820, when an English lawyer named Bigge began an inquiry on colonial administration, he gave evidence that helped to remove these imposts.

He also knew that although competition could be healthy a little co-operation, especially if it was rewarded, could also bring results. In 1825, he was largely instrumental in the formation of the Derwent Whaling Club, which in 1826 had a standing advertisement in the *Hobart Town Gazette*: "Derwent Whaling Club, 1826. Members Jas. Kelly, William Wilson, Walter Angus Bethune, and Charles Ross Nairne. Prize of 8 dollars is given to the first person who gives information of a whale being in the river. The profits of the club are divided into 7 shares; 5 are shared by the members, 1 to charitable purposes, and 1 to the native youth who displays the greatest expertness as headsman."

Although Kelly and the Derwent Whaling Club met with moderate success, there was something ominous in the appearance of the advertisement and the fact that a man could win a prize for sighting a whale. But for the time being all was well on the Derwent, and £74,000 worth of whale-oil and whalebone was exported to European markets from Tasmania in 1836.

It has often been suggested that bay whaling was an Australian invention, but it was practised by the Basques who fished along the French coast and in the Bay of Biscay in the twelfth and thirteenth centuries.

The Japanese also have a place in the history of shore-based whaling. The publication *Japanese Whaling Industry* (1954) says that:

Although the history of whaling in Japan dates back more than a thousand years, it was not until the early Seventeenth Century that whaling was established as an enterprise. Taichi Town of Wakayama Prefecture was the birth-place of commercial whaling. The descendants of the war-defeated feudal lord Wada who had emigrated and settled in the town began whaling for a livelihood. Through devices and improvements of the whaling method through the generations, the Wada family completed a commercial type of whaling which was later employed widely in Japan. The catch was made by a 'hand harpoon' method with spears, halberds, and arrows. This method, however, was replaced by 'net catching' in the late Seventeenth Century. As the whaling enterprise in these days consisted of catching operation, ship-building, black-smithing for gear, rope-making, oil and meat processing, hundreds of people were employed for one enterprise.

The feudal lords of the coastal regions encouraged and fostered whaling not only as a source of revenue but also as a reserve naval force. Since the invention of the netting method, commercial whaling spread throughout Western Japan and culminated in the early Nineteenth Century when whaling was carried on from more than thirty coastal bases in Japan.

A fleet for netting operation consisted of from eight to more than a dozen boats. Each boat was of about eleven meters in length, and two and a half meters in breadth with eight oars. Each boat was manned with about fifteen including a skipper, oar-men, and shooters. The boat was painted with various colours in order to be recognised from a long distance.

Headquarters was situated on a hill-top where a watch was kept over the sea. Several boats were also dispatched to find a whale. When a whale was found either by the sentry on the hill or by a scout-boat, a signal was made by means of smoke or a flag. The headquarters, after observing the current, wind and the speed of the running whale, gave instructions to catcher-boats by signals. Several catcher-boats drove the whale towards the shore. The men on the pursuing boat tapped the boat-side with hammers in order to frighten the whale.

When the whale set itself on a desired direction, the netting-boats lowered the net encircling the whale. Then catcher-boats came to a distance of about eight meters from the whale and threw harpoons to which ropes were attached.

These harpoons were not for killing the whale but to trace the direction of the whale with the attached rope when it submerged. When the movement of the whale was restricted by the net, heavier harpoons and double-edged

The South Australia Company's whaling station in Encounter Bay, in 1838

Colonel William Light's sketch-chart of the Encounter Bay area, site of South Australia's whaling stations

Chart of the Anchorages in Encounter Bay
by Wm. Light
Surveyor General

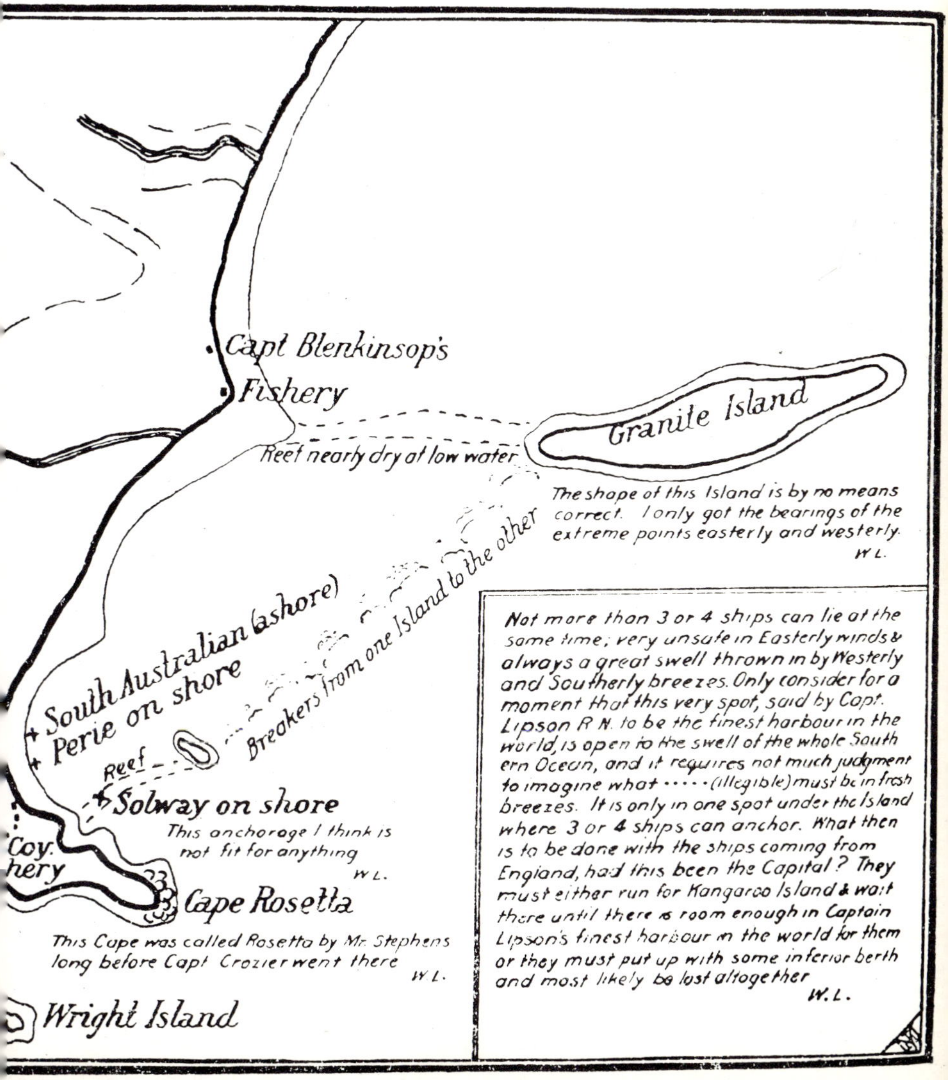

swords were used. The whale being almost killed had to be fastened to the tugging boats before it sunk. A man swam to the whale made a hole with a knife through the nose of the whale which was still alive, so this operation required the utmost care and courage. Another swimmer passed a rope through the hole. A third swimmer rode on the back of the whale and made a hole through the dorsal fin.

The whale was then further fastened with rope at various parts of the body and tied to a wooden bar held by two boats. Finally the finishing stroke was made with a sword. The whale rushed about in a death agony. At this point all the crew jumped into the sea to swim away. The dead whale was towed by boats to the shore and pulled up ashore with a pulley.

The whales were mostly utilized for human food. The blubber, cartilage and various viscera were eaten as well as the flesh.

As animal food such as beef and pork were not eaten in these days because of religious reasons, whale meat was highly appreciated. Various ways of cooking whale meat, therefore, were developed. *Whale Cookery*, published in 1780, describes a hundred and twenty recipes for whale meat. Whale-oil was mainly used for lighting but it was also used as insecticide for rice-plants. Bone and deteriorated meat were made into fertilizer.

Although the Australian may not have pioneered bay whaling, and the dangers could not be compared with the Japanese method, it is true to say that as an Australian industry it invented its own methods to suit the locality, framed its own laws, and certainly, along with deep sea whaling, produced a strain of men unique in Australian history: "Currency lads partial to the sport."

The "moment of truth" for the whaleboat crew came when the headsman delivered the harpoon. Sometimes the whale took an immediate revenge, as shown in the lower picture of a sperm whale attacking a boat (*Mitchell Library*)

4

Operations not attended with smell

IT WOULD BE AS INCORRECT TO INFER THAT ALL AUSTRALIAN whalers were rum-sodden loafers, who spent their watches below sharpening knives to cut each others throats, as it would be to say that all our pioneers were hard-working, moral specimens who would raise a slab church wherever three or four homesteads stood within walking distance of each other.

Whaling, like all industries, brought together a cross section of humanity; captains who came up through the ranks with greater skill in harpooning than text book navigation, young adventurers, spurred on by nothing more than too many red corpuscles in the blood, criminals on the run, and natural-born seamen who cared little for a ship's destination as long as it took them away from land.

But whaling had one big difference from all other industries. It was unique in that it called for its employees to venture forth in a small boat in mid-ocean and there come within striking distance of a monster which was likely to explode into gargantuan fury. In *Moby Dick*, Herman Melville spoke feelingly of the whalerman's possible fate when he wrote: "Do you suppose that that poor fellow there, who this moment, perhaps, caught by the whale line off the coast of New Guinea, is being carried down to the bottom of the sea by the sounding leviathan—do you suppose that that poor fellow's name will appear in the newspaper you will read tomorrow at your breakfast? NO: because the mails are very irregular between here and New Guinea. Yet I will tell you that upon one particular voyage which I made to the Pacific, among many others, we spoke thirty different ships, each of which had had a death by a whale, some of them more than one, and three that had each lost a boat's crew. For God's sake, be economical with your lamps and candles! Not a gallon you burn but at least one drop of a man's blood was spilled for it."

So there is no doubt that it required a special kind of fibre to ship aboard a whaler, and if a man did not possess it then he did not last very long. The French, British, American and Australian whalers which sailed out of Port Jackson and Hobart Town carried many different types of men, but they all had one thing in common. The ship's hold was empty; they were not carrying cargo from A to B, it had to be produced *en route* and delivered in a marketable form. And as Melville points out, mails were slow in those days. If they were mentioned in a newspaper it would only be when they returned to port, and then it would only be a comment on the size of the cargo—not the fact that men had died to light the lamps of peaceful citizens.

As the whalers lay alongside the wharves in Port Jackson and Hobart Town, the crew was coaxed aboard by smooth-tongued mates with promises of "home comforts," "the best and cheapest slop chest in the Pacific," and "enough grease to set yourself up for life." And if they had never been to sea before, they took an amazed glance around the deck as they lined up to sign their articles; tubs of rope, water casks, oil barrels, lines, harpoons, provisions, a pen containing a pregnant pig, and another with five or six hens; the captains usually had bacon and eggs for breakfast as long as they could.

Once they had made their mark they were committed to anything up to a twelve months' voyage if local grounds were favourable, and a three year voyage if the captain decided to sail further afield.

The first job was to clear the decks. Every item of gear had a special place so that mates could produce what was wanted at a minute's notice. Then the boat gear was checked; a boathook, hatchets, knives, drogues, tubs, mast, lugsail and boom, harpoons, lances and line, water breaker and emergency provisions. When fully loaded, they weighed up to half a ton.

Next, the boat crews were chosen. The ship's mates acted as headsmen, in charge of the boats; men with previous experience would be rated as harpooners; strong, fit seamen were chosen as pulling hands. It was most unusual for a whaler to sail with the same crew from one voyage to the next, and the captain was faced with the task of welding his men into efficient whaleboat crews within a few days of sailing. This was done with practice runs. One minute they would be going about their usual shipboard business, the next minute the alarm was raised and everyone raced for the boats; "hoist and lower" a dozen times a day till they could do it in the dark.

Meanwhile captain and headsmen worked out their ship's signals. Whales could be seen much better from the ship's masthead than from boats on the water, whose range of vision was limited, and an ingenious method of signalling using the ship's flags and sails was always worked out to guide them to their prey. For whales ahead, the jib was to be raised and lowered; whales to the windward, two flags at the mizzen; whales under the lee, three flags at the mizzen; whales between the ship and the boats, masthead flag up and down and half mast; and so on. If another ship put in an appearance the signals could be reversed by a pre-arranged plan; whaling captains had no intention of giving any advantage to an intruder.

Most whaleships were functional in design. They were three-masted, rather blunt in the bows, and cut off square at the stern, and they carried up to seven whaleboats slung over the bulwarks on heavy davits. The crew lived in a cramped fo'c'sle, entered by a companionway for'rard of the mainmast. Tiers of bunks filled the space from deck to deckhead, like wooden shelves into which the men fitted themselves to sleep on straw mattresses. These were supplied new at the beginning of the voyage, but soon became impregnated with the stench of rancid blubber. The harpooners, however, were a little better off, occupying quarters just for'rard of the stern cabins reserved for mates and the captain.

The only structure to break the flat deck, besides the hatch which opened to the hold, was the tryworks used for boiling down the blubber. Usually constructed of brick, it was fed first with wood, then with whale refuse, and housed two large cauldrons. One can imagine the ship rolling in a heavy swell, the flames licking up the sides of the cauldrons, canvas above flapping in the breeze, as the men struggled to cut the blubber into small pieces while others ladled the oil off into barrels. Many a whaleship caught alight during the hectic trying out period, and if rough weather was in the offing the men worked till they dropped rather than lose a whale secured alongside which may have taken them a day to kill.

For the first two or three weeks after leaving port, a whaler was usually a happy ship, but if no whales were sighted the fo'c'sle Jonahs took over. There was always someone on board who thought he knew more than the captain, and always some old hand who believed he could smell whales.

The captain became the scapegoat, and if the voyage lasted too long without any luck then he might be faced with a near-mutinous crew. Both strength of character and strength of arm were called for then, because the captain and his mates would have no option but to bully the crew into submission. They would start with a tongue-lashing, and if this failed them they would single out the ringleaders and go for them with fists, boots, and belaying pins. The captain made his own laws at times like these.

However, three magic words could transform a dry, scowling ship into a dedicated team: "Thar she blows," pealing from the masthead. Smoking pipes were thrust into trouser pockets, the captain's failings were forgotten, carving knives were dropped and arguments left unsettled as the crew raced for their boats. The mates yelled, "Clear away . . .

hoist and lower!" Davits creaked, men swore, and within minutes the decks were clear and the boats were in the water. If it was a large school, it was open slather. If it was a lone whale, the mates flung a glance at the ship's signals, for confirmation of the original direction of the sighting and ordered the sail to be set.

They were now in a world of their own. The ship receded with every stroke of the oars or filling of the sails and the headsman grasping the steering sweep became their eyes. Some chases were long, others surprisingly short. As far as vision was concerned, nature was on their side. Although sperm whales have sensitive hearing they cannot see ahead or astern, so they were generally approached from behind, but if possible clear of the tail. With their backs to the prey the men rowed automatically, stealing an occasional glance over their shoulders to reassure themselves that they were not being fooled by another practice run. If the weather was calm the oars were boated and paddles taken out to make the approach as quiet as possible.

At a command from the headsman, the harpooner left his oar. Bracing himself in the bow crotch, he raised the harpoon, took quick aim at the huge glistening bulk of the whale, and threw the heavy weapon with all his strength. The result was cataclysmic. The sperm could either be drowsing on the surface or preparing to dive for squid, but whatever his mental condition, the sudden agonising entrance of the harpoon into his flesh had only one effect. He was immediately geared for survival.

One of the best descriptions as to what follows is to be found in the writings of Thomas Beale, demonstrator of Anatomy to the Electric Society of London and surgeon of the *Sarah and Elizabeth*, which made a whaling voyage to the Australian grounds in 1838.

The glistening harpoon is seen above the head of the harpooneer, in an instant it is darted with unerring force and aim, and is buried deeply in the side of the huge animal. It is "socket up," that is, it is buried in his flesh up to the socket which admits the handle or pole of the harpoon. A cheer from those in the boats and from the seamen on board, reverberates along the still deep at the same moment. The sea, which a moment before was unruffled, now becomes lashed into foam by the immense strength of the wounded whale, who with his vast tail strikes in all directions at his enemies. Now his enormous head rises high into the air, then his flukes are seen lashing everywhere, his huge body writhes in violent contortions from the agony the "iron" has inflicted. The water all around him is a mass of foam, some of it darts to a considerable height—the sounds of the blows from his tail on the surface of the sea can be heard for miles. "Stern all," cries the headsman; but the whale suddenly disappears; he has "sounded"; the line is running through the groove at the head of the boat with lightning-like velocity; it smokes—it ignites, from the heat produced by the friction, but the headsman, cool and collected, pours water upon it as it passes. But an oar is now held up in their boat; it signifies that their line is rapidly running out; 200 fathoms are nearly exhausted; up flies one of the other boats and "bends on" another line, just in time to save that which was nearly lost. But still the monster descends; he is seeking to rid himself of his enemies by descending deeply into the dark and unknown depths of the vast ocean. They next bend on the "drogues" to retard his career—but he does not turn; another and another have but slight influence in checking the force of his descent; two more lines are exhausted—he is six hundred fathoms deep. "Stand ready to bend on," cries the mate to the fourth boat (for sometimes, though not often, they take the whole four lines away with them—800 fathoms), but it is not required, he is rising; "Haul in the slack," observes the headsman, while the boatsteerer coils it again carefully into the tubs as it is drawn up. The whale is now seen approaching the surface; the gurgling and bubbling water which rises before also proclaims that he is near; his nose starts from the sea; the rushing spout is projected high and suddenly, from his agitation. The "slack" of the line is now coiled in the tubs, and those in the "fast" boat haul themselves gently towards the whale; the boatsteerer places the headsman close to the fin of the trembling animal, who immediately buries his long lance in the vitals of the leviathan, while at the same moment, those in the other boats dart another harpoon into his opposite side, when "stern all" is again vociferated, and the boats shoot rapidly away from the danger.

Mad with the agony which he endures from these fresh attacks, the infuriated "sea beast" rolls over and over, and coils an amazing length of line around him; he rears his enormous head, and, with wide expanded jaw, snaps at everything around; he rushes at the boats with his head— they are propelled before him with vast swiftness, and sometimes utterly destroyed. He is lanced again, when his pain appears more than he can bear; he throws himself, in his agony, completely out of his element; the boats are violently jerked, by which one of the lines is snapped asunder; at the same time the other boat is upset and its crew are swimming for their lives. The whale is now free; he passes along the surface with remarkable swiftness, "going head out," but the two boats that have not yet "fastened," and are fresh and free, now give chase; the whale becomes exhausted, from the blood which flows from his deep and dangerous wounds, and the 200 fathoms of line belonging to the overturned boat which he is dragging after him through the water, checks him in his course; his pursuers again overtake him, and another harpoon is darted and buried deeply in his flesh. The men who were upset, now right their own boat without assistance from the others, by merely clinging on one side of her, by which she is turned over, while one of them gets inside and bales out the water rapidly with his hat, by which their boat is freed, and she is soon again seen in the chase.

The fatal lance is at length given—the blood gushes from the nostrils of the unfortunate animal in a thick black stream, which stains the clear blue water of the ocean to a considerable distance around the scene of the affray. In its struggles the blood from the nostrils is frequently thrown upon the men in the boats, who glory in its show. The immense creature may now again endeavour to "sound" to escape from his unrelenting pursuers; but it is powerless—it soon rises to the surface and passes slowly along until the death pang seizes it, when its appearance is awful in the extreme. Suffering from suffocation, or some other stoppage of some important organ, the whole strength of its enormous frame is set in motion for a few seconds, when his convulsions throw him into a hundred different contortions of the most violent description, by which the sea is beaten into a foam, and boats are sometimes crushed to stems, with their crews.

But this violent action being soon over, the now unconscious animal passes rapidly along, describing in his rapid course a segment of a circle; this is his "flurry" which ends in his sudden dissolution. And then the mighty rencontre is finished by the gigantic animal rolling over on its side and floating, an inanimate mass, on the surface of the crystal deep—a victim to the tyranny and selfishness, as well as a wonderful proof of the great power of the mind of men.

In Beale's description the harpooner who is standing in the bow of the boat throws the harpoon, but later the headsman, who began in the stern acting as boat-steerer, is to be found in the bows. This is not a mixture of terms. These two men were acting out one of the most curious customs of the sea known in whaling as "the change." Beale says that: "The crew of the boat consists of the headsman, boat-steerer, and four hands; of these, the headsman, who has the command of the boat, steers it until the whale is reached and struck with the harpoon. The boat-steerer also, at this time, pulls the 'bow oar,' but when near the whale he ceases rowing, quits the oar and strikes the harpoon into the animal, the line attached to which runs between the men to the after part of the boat, and after passing two or three times round the loggerhead, is continuous with the coils lying in the tubs at the bottom of the boat. The boat-steerer now comes aft, and steers the boat by means of an oar passed through a ring attached to the stern called a 'grummet'; he also attends the line through all the subsequent operations; the headsman at the same time passes forwards, and takes the station at the head of the boat, prepared to plunge his lance into the body of the whale at the first opportunity, and it requires considerable tact and experience to do this in the most effectual manner."

Why they chose to do this at such a crucial time in the operation is not known. There must have been enough confusion and danger without two men actually changing places and passing each other amidships as the boat was being towed through the water. A man who could throw a harpoon into a whale could surely use a lance to penetrate its lungs, or "find its life" as the whalers said, but possibly at some time in the formative period of deep sea whaling it was decided that the headsman, being in command of the boat, should have the honour of making the kill. This little manoeuvre, which was also practised by Australian whalers, certainly did not point to the "great power of the mind of men."

After the death of the whale the men only had a few minutes respite. They were now in possession of a valuable carcass, and whatever their physical condition it had to be towed back to the ship before it sank. This could mean several hours of rowing, depending on how far they had been towed during the fracas, and possibly against the wind or in rough sea. Even two hundred yards must have seemed like a mile to men who had undergone the tremendous physical and emotional strain of the kill.

Beale then goes on to describe the next manoeuvre:

> After the death of the whale, the next steps are to remove the fat and spermaceti and to extract the oil, the reward of so much exertion and dangerous enterprise. This process, which is called "cutting in," is divided into two stages, the cutting in and trying out—in these operations the utmost cleanliness is observed. As soon as possible after the whale has been killed, it is brought alongside the ship to be cut in, by means of instruments which are called "spades."
>
> A man descends upon the floating carcass, after a hole has been cut near the junction of the body with the head

by a person stationed on a stage (platform) by the side of the ship, into which a large blunt hook is inserted by the man who descends upon the body of the whale for that purpose, and which is called "hooking on," a troublesome and even dangerous operation in a turbulent sea, from which I have often seen men drawn up by the rope which is always fixed around them, in a completely exhausted state, from having been frequently submerged by the waves as they have rolled over the body of the whale during the time they have been endeavouring to "hook on." After the hook is inserted it is drawn upon by the pullies to which it is attached, and a tension being created upon the fat, or blubber as it is termed, it is then cut by the spade in a strip about two or three feet broad, in a spiral direction around the body of the whale, which being drawn up by means of the windlass acting upon the pullies which are fixed to the "main top" it is removed much in the same way that a spiral roller or bandage might be; of course as the "blanket pieces" ascend the body of the whale performs a rotary motion until the whale is stripped off to the tail or flukes. The head is cut off in the beginning of the process and is allowed to float astern of the ship, but strongly secured, until the body has undergone its flaying process, when it is hoisted upon end by the pullies, and the end of the case being opened, the fluid spermaceti is drawn from it by means of a bucket and pole, which is used to force the bucket down into the "case," so as to become filled with its valuable contents. The "junk," which forms, when it is cut from the head, a large wedge-shaped mass of cellular substance, having strong ligamentous bands intersecting and strengthening its structure, is hoisted on board, and quickly cut up into square oblong pieces, while the head—after the case has been emptied—is let go, and allowed to sink, which it does rapidly when its buoyant property, the spermaceti, is removed.

The lean and shapeless trunk, after the fat has been taken from it, is allowed to float away just before the operations on the head commence. The "blanket pieces" which are cut from the long strips of fat or blubber, as it is drawn up, with the junk and spermaceti from the case, then pass through different processes in the "trying out"; the two former, being cut up into thin slices upon blocks called "horses," are thrown into the "try pots," in which the oil is extracted from them by heat. The crisp membranous parts after the oil is extracted, and which are called by whalers "scraps," serving for fuel; while the spermaceti from the case is carefully boiled alone, and placed in separate casks, when it is called "head matter."

These operations are not attended with any unpleasant smell, and are very quickly performed, as eighty barrels of oil may be stowed away in the hold of the ship in less than three days after the destruction of the animal.

The last remark about the operation not being attended by any unpleasant smèll is most unusual. Other old whalers describe "cutting in" and "trying out" as "the purgatory on deck," of men "barefooted in fat and blood," and complain bitterly of the "sickening reek in their nostrils." Perhaps Beale turned a blind nostril to it and described it from a distance.

Apart from "the change" Australian whalers also adopted another traditional rule of whaling in the method of lay payment, a lay being a share of the profit from the voyage. It was very convenient for the owners, as it relieved them of the responsibility of paying any wages if the voyage was unprofitable. In 1834, James Kelly, owner of the brig *Mary and Elizabeth*, paid the following: seamen one-fortieth to one-sixtysecond, headsman one-eleventh, cook and steward one-sixtieth. The captain, however, usually received one-eighth while mere deckhands could be as low as one-seventieth. A small share was called a long lay, and a big share, such as the captain's, a short lay.

On paper this seemed fair enough, especially when some ships returned with £5,000 worth of oil, but the cards were definitely stacked against the seamen. They were fed as part of their agreement, but as sailors they needed soap, tobacco, knives, and a few luxuries of life. These they could buy from the ship's slop chest (or diddle box as it should have been called) at a price. When the ship arrived home the seamen were told the amount of their lays, but immediately the value of the goods they purchased from the slop chest was deducted. A clean, heavy-smoking pulling hand could find himself well nigh in the red at the final reckoning, and would be lucky if he cleared a few pounds.

Deep-sea whaling was a low-paid, dangerous trade, and apart from the paid game of war and some sports, it would be hard to compare it with any modern occupation.

It is easy to imagine our whalers as tough, blaspheming hard-drinking individuals who were denied skilled jobs ashore; forced to take to the sea to earn a living. But many of them had wives and children in Hobart Town and Port Jackson and no doubt in different circumstances, depending on their individual character, showed the usual affection in the domestic environment. Like many men of the sea, they were forced to accept the loneliness of an all-male situation, and no doubt split their personalities; reserving one part for their working day with its cursing and swearing and back-breaking tasks, and the other for the quietness of the watch below, dwelling on more pleasant memories as they carved trinkets out of whale-teeth, not to pass the time but as gifts for their children.

In many cases they were strangers in a new country and were forced to accept its challenges for survival; forced to provide for their families. There was no turning back, and since whaling was an established industry many of them took to it merely for bread and butter. Once arrived in Australia they were, like most of our pioneers, victims of circumstance. Their only concern was to earn a living for themselves and their families, and what may seem to us to be a picturesque chapter of history was to them merely an everyday occupation.

But in every situation there are a privileged few, and in deep sea whaling the captain was the most prominent of these. He enjoyed the best available food, was in a position to make a few pounds on the side by juggling the slop-chest prices, and if his marital condition was equal to it, he could take his wife on the voyage. Even if he sailed alone, his cabin boy was available to clean his shoes, dry his clothes, and serve hot drinks at any time.

Few men of the sea, however, had the responsibility that went with the command of a whaler. A great deal depended on the captain's knowledge of whaling grounds; he had to keep his grease-hungry crew happy, and satisfy the owners on his return. Nowhere else in maritime history do we find the captain leaving his ship in mid-ocean to assist in the production of a cargo *en route*. As a captain he was expected to be a business-man, have a nose for grease, and to be a skilled whaler. Unlike the men who served under him, the captain of a whaler could only afford the luxury of more pleasant meditation when his ship was grease from stem to stern.

5

No employment more disgusting

"I EMBARKED ON BOARD THE COMPANY'S BRIG *Emma* BOUND TO Encounter Bay, with a view to witnessing the whale fishery. After a pleasant sail from Kingscote (on Kangaroo Island) of twenty-four hours we cast anchor in the Bay. The scenery along the shore was one of unvaried wilderness, with a bold iron-bound coast, to which a vessel may sail so close as to strike her studding-sail boom and yet be in deep water. As we got nearer Encounter Bay the view at once, as if by magic, assumed a more cheerful appearance, opening into a wild but peaceful park, which reminded one of the domain of an English noble. In the bay was riding the little craft that brought me so many thousand miles; there she lay, but how sadly altered. No rigging, nothing but her stumps; a perfect model still, smiling through her grease and wretchedness. Her once spruce, bluejacket tars were now leaning over her gunwhale and looking as filthy and as savage as the untutored barbarians around them."

The little craft that Dr W. H. Leigh referred to in his writings in 1840, *Voyages, Travels and Adventures*, was the *South Australia*, on which he had sailed from England. She was part of the fleet of the South Australian Company, which was formed in London in 1836 to set up, among other enterprises, whale fisheries along the South Australian coastline.

On the banks of the River Derwent, bay whaling grew out of a penal colony, but in South Australia it was to be pioneered by a group of investors 12,000 miles away, who had been encouraged to part with their money by the glowing reports of profits from ships arriving in London from Hobart Town.

Although he was impressed by the natural beauty of the bay, the doctor was appalled by the condition of the ship and her crew, and the stench of the tryworks and rotting whale carcasses was too much for his clinical nostrils. He was further moved to write: "There is no employment more hazardous, more laborious, more disgusting than whaling. The station-barrack, or house inhabited by the whalers on shore, was a kind of barn where these poor fellows indiscriminately turned in, self-exiled from the society of all human beings but their own few, living among savages as wild and reckless as they."

As far as the whalers themselves were concerned he was right. Isolated as it was, Encounter Bay made a perfect haven for the flotsam and jetsam of the industry; escaped convicts, ships deserters, and others who for various reasons wanted to escape from civilization, made it their home. In his *Reminiscences*, Alexander Tolmer, South Australia's first Commissioner of Police, complained that rum and tobacco smuggled ashore from American whalers in the bays and inlets of the south coast eventually found their way to Adelaide. No doubt the Encounter Bay whalers could have helped him in his enquiries, but whenever he or anyone else with law enforcement authority visited the area they were either out to sea, inland cutting timber, or on their best behaviour.

One wonders if the shareholders in London would have been so ready to invest their savings had they known the type of men responsible for turning the raw material of the industry into profit. And yet it cannot be denied that they *were* men. Rum may have given them the "bravery of fools," but muscle and stamina were needed for the chase, throwing skill for the

The agonized upheaval of a harpooned whale was known, perhaps with grim humour, as "the flurry" (*Mitchell Library*)

The cutting-in, showing a whaler pulling a long strip of blubber from a whale by a block and tackle on the foreyard (*Mitchell Library*)

harpoon and lance, and agility to save legs as the line rattled out after a strike.

Although faced with the usual difficulties of a distant market, and an area which attracted more than the usual quota of undesirable employees, the South Australian Company was fortunate in the natural assets of their bay whaling site.

Lying fifty miles south of Adelaide, it extends in a sweeping curve from Newland Head at the western end to Cape Jaffa in the east. It was named by Captain Matthew Flinders to commemorate his encounter with the French Captain Nicholas Baudin on 8 April 1802, about eleven miles south east of the Murray mouth.

The western end of the bay attracted the whalers, and in particular Rosetta Head, or the Bluff as they called it, rising some 320 feet above sea level.

Except for a memorial plaque outlining the meeting of the two navigators, the granite-strewn peak, almost bare of vegetation, has changed little in the last 130 years. In a gale, the wind still howls like a thousand devils, and the rocks which no doubt hid the secret caches of the lookout's rum or whisky are as immovable and cold as when the whalers climbed them to scan the seas at sunrise.

Apart from the Bluff, four small islands lying close in shore gave protection from southerly gales. The largest of these, Granite Island, opposite to where the town of Victor Harbor now stands, provided a sound, safe anchorage.

Captain John Hindmarsh, South Australia's first governor and a fanatical advocate for Encounter Bay as the site of the capital, saw the possibility of the Bluff. He was allotted the section of land on which it stood. If he had won the argument with Colonel William Light, Surveyor-General for the colony, and others who favoured the present site of Adelaide on the Gulf of St Vincent, he would have had the most commanding situation for a residence in the colony.

After his recall to England, the section passed to his son John, and in June 1855, when the local district council began work on a road skirting the edge of the Bluff to a new screw-pile wharf, John Hindmarsh Junior sued them for £48,000

From a whale secured alongside, chunks of blubber have been hacked off and are being hoisted aboard for trying-out (*Mitchell Library*)

These whalemen are preparing to split the "case;" the organ in a sperm whale's head which contains many gallons of pure oil (*Mitchell Library*)

for encroaching on his land. This resulted in the insolvency
of the Inman District Council, a lengthy litigation, a final
settlement of £2,000 in 1859, and the South Australian
Government assuming control of all the land in the
Hindmarsh section.

Although Encounter Bay was known to whalers and sealers
as early as 1802, it was not until March 1837 that permanent
whalers arrived at the Bay. In that month a party from
Sydney in the *Hind*, under Captain Blenkinsopp, and a
double party of six boats from the South Australian Company, under Samuel Stephens, arrived within a few days
of each other.

Captain Blenkinsopp, on behalf of Robert Campbell
Junior and Company, a firm of Sydney merchants, set up
camp about half-a-mile north of the present causeway
connecting Granite Island and the mainland, but later, no
doubt because of the need of high land for a lookout, shifted
his camp to the island.

Blenkinsopp must have had great faith in the future of the
industry as he lost no time in building a home, which he
named "Anne Vale," after his wife, near the present
Cornhill Road. He also named the area Hind Cove, but
the name never appeared in any official records.

Meanwhile, Samuel Stephens had chosen a small inlet
at the foot of the Bluff for his station.

Encounter Bay whalers, described by contemporary
writers as the "scum of the earth" and "rum-soaked loafers,"
must have had their moments. Over the years, Stephens'
men blasted out launching ramps in the granite rocks on the
shoreline. Evidence of this can be seen from the road now
running around the foot of the Bluff to the screw-pile jetty.

They also hacked away at the face of the Bluff to make
room for their try-works and barrelled oil. But the area in
which they worked was cramped compared with Blenkinsopp's station on Granite Island, which had a natural sandy
beach for launching the boats and ample flat land for
storage.

Both parties evidently considered that shelter from the
winter gales, which they both had in the lee of the island and

the Bluff, and high land for lookouts, was more important than space, as they seem to have ignored the open stretch of beach which runs eastward from Granite Island to the Nob, now known as Port Elliot.

Both parties were in a good position to launch the boats for the chase. Blenkinsopp's men could skirt the seaward tip of Granite Island, and race across the bay to intercept the whales as they headed inshore, while Stephen's men, two miles to the west, could pull out to meet them head on.

Arriving in March, both parties should have been ready for the season beginning with the onset of winter. In their report of 20 May 1836, the directors of the South Australian Company said: "Your Directors are promoting the establishment of off-shore stations for catching the black whales, amidst the bays of the colony, and they hope to be ready for the next season and obtain profitable return."

Having definitely committed itself and its shareholders to the wavering fortunes of bay whaling, the Company waited for the annual visitors to come to Encounter Bay in the winter of 1837.

The following, taken from a report of the Statistical Society of South Australia in 1841, explains the migration habits of whales at the time:

> The general course of the black whale in these seas, as winter approaches, appears to be from the south-east, consequently the southern shore of Van Dieman's [sic] Land is first visited by them, which may be about the beginning of April. These move on towards Portland Bay; others continue through the winter to arrive and pass forward. Of those which enter Encounter Bay some have probably coasted along from Portland Bay, while others, it would appear, strike the coast there for the first time. In like manner the whole southern coast of this continent is visited by them, some having come along the land, whilst others are more direct from the great Southern Ocean. At Cape Lewin the great body of whales seem to strike off southward, for in October and November they are again making towards the south-east, by keeping two or three

hundred miles from the land, where they are again pursued by vessels engaged in the 'Off-shore Fishery'.

The black whale mentioned here is the right whale, which was black in colour.

Waiting for the whales in a closed bay, with two rival companies, two look-outs, and the men on small lays, created an explosive situation. Captain Blenkinsopp sensed that he was in the calm before the storm, and approached Samuel Stephens with an offer for the parties to work together, as was customary in New Zealand. But Samuel Stephens refused, and the stage was set for a battle which came close to ruining the chances of both companies.

Despite this, Captain Blenkinsopp's party managed to secure 200 tons of whale oil which he claimed was the first commodity exported from the infant colony of South Australia. On the other side of the bay, Samuel Stephens managed 160 tons of oil and a quantity of whalebone which, when shipped to London in the *Seppings* netted him £25 and £100 a ton respectively.

By June 1837, however, the whalers were at each others' throats, Captain Blenkinsopp complaining that the South Australian Company was stealing his men by offering them higher lays, and Samuel Stephens complaining that Blenkinsopp was interfering in his fishery. When all but 10 of his men were lured away by his rival, Captain Blenkinsopp threatened legal action and claimed compensation of £11,000 for loss of oil. Stephens wanted Blenkinsopp ousted from the Bay, and Blenkinsopp wanted his men back.

Into this explosive muddle rode the South Australian Resident Commissioner, James Hurtle Fisher, who set out from Adelaide with Colonel William Light as a magistrate. But they were held up by floods, and it was two months before the complaint was heard. Colonel Light decided he had no jurisdiction to oust Captain Blenkinsopp, and the matter lapsed into an uneasy truce.

Although the South Australian Company had won the first round, they were soon plagued by domestic trouble.

The double party which took 160 tons of oil was overpaid.

Before paying the men, the quantity of oil should have been divided into two, and the lays of each party worked out upon one half of the total. As it was, Stephens worked out each man's share on the whole catch, with the result that the lays totalled up to more than the value of the oil received.

This arithmetical blunder and the dispute with Captain Blenkinsopp, however, were not the South Australian Company's only worries. On 8 December 1837, the Company barque *South Australia* ran aground on the east side of the Bluff during a violent gale and became a total loss. This disaster was followed later by the wreck of two other Company ships in the same place, the *John Pirie* and the *Solway*.

In reporting these losses to the shareholders, Mr George Fife Angas, Chairman of the Company, hastened to inform them that "all vessels had earned a considerable amount of freight and passage money on the outward voyage" and that "the insurance on these vessels will reimburse the Company for their cost and outfit and thereby preserve you from pecuniary loss."

Despite this, it must have been a bitter blow to the whaling hopes of the Company at Encounter Bay, and one which could possibly have been avoided, for while Colonel Light was in Encounter Bay he passed the time between hearing evidence in practising his profession of surveying and drew a chart of the area. Of the anchorage in the lee of the Bluff, used by the South Australian Company, he said: "This anchorage I think is not fit for anything." His recommendations were heeded in August 1838, when the *Goshawk* dropped anchor in the lee of Granite Island, instead of the Bluff, to load the South Australian Company's whalebone and oil.

The second season at Encounter Bay (1838) brought changes. Captain J. Hart was placed in charge of the South Australian Company's station, and a Mr J. B. Hack bought the Granite Island station from Captain Blenkinsopp. Hart was an experienced whaler and sealer, and had made three voyages from Launceston along the South Australian coastline between 1831 and 1834.

At this stage, Captain Blenkinsopp, pioneer whaler of South Australia, came to an untimely end. At the time of the great controversy between Colonel William Light and Governor Hindmarsh over the site of the capital, a drunken whaler from Kangaroo Island named Walker, turned up in Adelaide, with the claim that he knew of a deep harbour at Encounter Bay with a channel leading to the sea.

When challenged, he asserted that it was not the channel which Charles Sturt had found, and the one subsequently examined by Captain Collet Barker and later by Colonel Light, but another, twenty-five miles south-east of Sturt's channel running into Lake Alexandrina.

Walker claimed he had seen it before any of the explorers, while he was on "gin raids" to the native camps around the Murray mouth. The news not only rekindled the dispute between the two chief antagonists, but led to Governor Hindmarsh asking Sir John Jeffcott, in company with Captain Blenkinsopp, to re-examine the area. If Hindmarsh could have proved the existence of a deep sea harbour anywhere near the Murray mouth, then he would have won the day and caused Adelaide to be shifted to Encounter Bay—no doubt with increased support from the settlers.

Unfortunately, Captain Blenkinsopp, a whaler to the last, took aboard a head of whalebone he found on the beach. The boat was swamped and sank in the surf of the Murray mouth. Blenkinsopp's body was the only one recovered, and he was buried in the garden of his home at Anne Vale.

During this second season, experienced whalers seemed to be scarce. In his report, Captain Hart maintained that out of a four-boat party belonging to Hack's station only two headsmen were capable of killing a whale. Evidently as long as a man could pull an oar he could claim to be a whaler. Despite this, they managed to take 104 tons of oil.

In 1839, the South Australian Company and Hack & Company joined their interests under one manager, but still many of the men recruited had no previous whaling experience.

The South Australian Company had also established stations at Thistle Island at the foot of Spencer Gulf, and at Sleaford Bay on the tip of Eyre Peninsula, but when the whales did not appear the boat crews were transferred to

Encounter Bay, where the pulling hands deserted to a man. Despite this, 176 tons of oil were taken during the season. Much of this was lost, however, when the *Katherine Stewart Forbes*, chartered to load in July, did not drop anchor in the bay until January.

The Company attempted to regain the loss by legal action, but this was defeated when by a legal error they sued the captain instead of the owners.

These losses, and the disappointments they had suffered during the first three years of whaling, moved the South Australian Company to a programme of gradual withdrawal from bay whaling, and they began disposing of their assets.

From this it is clear that whaling in Australian waters was a risky business, not only for the men in the boats, but also for those responsible for the Profit and Loss Account.

The following account, printed by The Statistical Society of South Australia in 1841, reveals the method of calculating net profit on a catch of only 100 tuns of oil and five tons of bone. (Whale-oil was always reckoned in tuns, an old English liquid measure of seven Colonial barrels each holding approximately thirty gallons of oil).

The men are paid by "lays" (or shares), which are certain proportions of the oil and bone caught, and reckoned at a price agreed upon at the commencement of the season, which lays and prices vary according to the rank and skill of the respective men. The lay of the chief headsman, who has control of the whole party, varies according to his character and skill, from the 11th to the 13th. The second headsman has from the 15th to the 17th; the third has from the 17th to the 20th. To "young headsmen" or those who for the first time act in that capacity, the 25th lay is given. Boat-steerers have the 40th, and the pulling hands the 60th to the 65th. The cooper has £8 per month. This will, however, be rendered more intelligible by a calculation of the lays of a three-boat party, reckoned upon one hundred tuns of oil and five tons of bone. . . .

(See overleaf for table drawn up by the Statistical Society.)

EXPENSES AND PROFIT ON A SMALL "CATCH" OF 100 TUNS OIL, AND FIVE TONS WHALEBONE

	Lay	Price of Oil	Value	Price of Bone	Value	Total Value	Proportion	£ s. d.	Amount of Lays	£ s. d.
1 Chief Headsman	13th	£10 per tun	£1,000	£65 per ton	£325	£1,325	1.13			100 8 4
1 Second Headsman	15th	£10 per tun	£1,000	£65 per ton	£325	£1,325	1.15			88 6 8
1 Third Headsman	18th	£10 per tun	£1,000	£65 per ton	£325	£1,325	1.18			73 12 3
3 Boat-steerers	40th	£7/10/– per tun	£750	£35 per ton	£185	£935	1.40	23 7 6 x 3 =		70 2 6
1 Steward	50th	£7/10/– per tun	£750	£35 per ton	£185	£935	1.50			18 14 0
1 Cook	60th	£7/10/– per tun	£750	£35 per ton	£185	£935	1.60			
1 Bullock driver	60th	£7/10/– per tun	£750	£35 per ton	£185	£935	1.60	15 11 8 x 19 =		295 1 8
17 Pulling hands	60th	£7/10/– per tun	£750	£35 per ton	£185	£935	1.60			
1 Cooper, 6 months at £8 per month										48 0 0

27 is the total number employed, the expense of whose labour is   £694 5 5

The provisions for the men are furnished by the proprietors on the following scale:

			£ s. d.
12 lb flour per man per week or for season of 25 weeks equal	300 lb at 27/– per 100 lb		4 1 0
12 lb meat per man per week or for season of 25 weeks equal	300 lb at 5½d per lb		6 17 6
¼ lb tea per man per week or for season of 25 weeks equal	6¼ lb at £10 per chest		0 12 6
2 lb sugar per man per week or for season of 25 weeks equal	50 lb at 3½ per lb		0 14 7
2 quarts spirits per man per week or for season of 25 weeks	equal 50 galls. at 3/– per gall. bd.		0 14 7

Provisions to one man for season is   £16 0 7

Or for the whole 27 hands employed   432 15 9

Add for leakage, draft, waste, etc., 7½ per cent   32 9 1

Small stores for headsmen and extra utensils   5 0 0

     470 4 10

The other charges may be stated as under, viz.:

Freights to the bays of stores, gear, etc., say   100 0 0

Wear and tear, one year with another   350 0 0

Contingencies   100 0 0

     550 0 0

Whole expenses exclusive of interest of money, etc.   1,714 10 3

100 tuns oil cannot be valued for exportation without casks at less than £15 per tun, or   1,500 0 0

5 tons bone   500 0 0

Total Value   £2,000 0 0

Net profit on catch of only 100 tuns oil and 5 tons bone   285 9 9

     £2,000 0 0

6

After the passing of this act

In september, 1839, john dutton, chief headsman of the Granite Island fishery, heard that some of his men, dissatisfied with his administration, had conspired to tap the oil casks and desert. He set a watch, and when two intruders could not give a satisfactory account of themselves, he placed them in chains and made arrangements with the captain of a schooner in the bay to take them to Adelaide. One, Alexander Riches, escaped, and while trying to cross the reef from the island to the mainland, fell in a hole and drowned. Dutton was arrested, and taken to Adelaide to face a charge which was solemnly read out to him as:

That John Dutton, not having the fear of God before his eyes, but being moved and seduced by the instigation of the Devil, did bruise, beat and otherwise illtreat Alexander Riches, and did put him in bodily fear of his life. The said Alexander Riches did then make his escape from the said John Dutton into a certain rock or pass, commencing at the point of the rock or island, called Granite Island, and connecting the same with the mainland. Alexander

Riches was thrown from the said rock or pass into the water and there choaked, suffocated, and drowned, of which said choaking, suffocating, and drowning, he the said Alexander Riches, then and there instantly died.

This first real attempt at law enforcement at the bay, however, proved a fiasco. The charge of manslaughter was dropped when witnesses for the prosecution failed to appear; the whalers evidently preferred to settle disputes in their own way. This they proceeded to do with a vengeance.

Some time later, Alexander Tolmer, Superintendent of Police, was called to the bay when Captain Lindsay of the barque *William the Fourth* lodged a complaint of assault by Isaac Clark, a whaler in the employ of Captain Hart. It appeared that when several boats were closing in on a whale, Clark's boat was a little ahead but being overtaken by Lindsay's. So Clark rammed him, threw a harpoon into the boat, and clouted one of the men with an oar. The arm of the law may have been long, but it was also costly. Clark was fined ten shillings—small reward for sending a party fifty miles south from Adelaide.

But what of the whales themselves during these mêlées? The report of the Statistical Society of South Australia dismisses them in a few lines: "Great competition will surround the whale by many boats, and when so disturbed he is apt to become enraged or galled and thus commit much damage among them."

The whales were not the only ones to become enraged. At times, as many as eight boats closed in on one target, and harpoons were often directed at a rival boat instead of at the whale.

The land-based whalers were also up against the faster and more modern boats from visiting whaling ships. In 1840, four French and American whalers dropped anchor offshore, and in the same year Captain Hart, manager of the whale fishery, wrote: "Unfortunately there was great opposition, two vessels being there at the same time from Van Diemen's Land. Now a new evil presented itself. The boats being built upon a new and improved model pulled much faster than those at that time used in this province. Not infrequently were the boats belonging to the shore parties within a few fathoms of the whale when the Van Diemen's Land boats would shoot past them and fasten to the fish and causing great dissatisfaction among them. Under these circumstances, although the evil could not have been seen or provided against, it was found impossible to limit them to a stipulated quantity of provisions. In ordinary years this would not have been of so much consequence, but with flour at from £50 to £80 per ton and pork at £7 per barrel, the wasteful extravagance that the owners could not stop made profit all but impossible."

It seemed that in a good season the whalers would work on normal rations, but drowned their frustration by gorging themselves in times of keen competition.

News of these mêlées reached Adelaide, and in May 1840 Captain Sturt and a strong reinforcement of police went down to restore order. As usual, when they arrived, the whalers were on their best behaviour and Sturt found that "there was practically no foundation for the rumours."

One can imagine the rum-inspired oaths echoing across the bay as the boats closed in for the strike; it would be so easy to settle old scores by a misplaced harpoon in the confusion of boats, high seas, and rattling lines. And what altercations must have taken place on the beach, while splintered boats and broken legs were mended, with no guiding principles of law or rights of possession of the fish to guide the antagonists.

But in 1840 a most valiant piece of legislation was nailed to the whalers' huts; "An Act for the regulation and protection of the Whale Fisheries."

After the passing of this Act, in every case where a whale shall be struck in any river, bay, or inlet of this Province, or on the high seas within twenty miles of the shore with a harpoon or any other instrument used in killing or taking of whales, the following shall be the rules for deciding upon the property in any such fish so struck (that is to say): First—Where the harpoon or instrument shall remain in the fish so struck, and a line and boat shall be attached thereto and continue in the power of the striker or headsman, such fish shall be deemed a fast fish, and every harpoon or instrument struck by any other person shall in such case be deemed a friendly harpoon, and the fish shall be the property only of the first striker or headsman. Secondly—In case the line affixed to the first harpoon or other instrument so struck shall break or the line attached thereto shall not be in the power of the striker, but the harpoon or instrument shall nevertheless remain fast in the flesh, then the fish taken or killed by any subsequent harpoon shall be the joint property of the first and second or subsequent striker in equal proportions. Thirdly—If the first iron or harpoon so struck as aforesaid shall break or shall become disengaged from the fish, then the fish shall be considered a loose fish.

The Act also provided that men's articles for service in the whale fisheries must be in writing, and that breach of their terms was to be an offence. This was included in an effort to stop desertions, which at times left the stations almost unattended except for headsmen.

This Act, although similar to one already observed by Tasmanian whalers, was evidently not worth the paper on which it was written. Captain Hart wrote that: "Although a good and stringent law had been passed for the regulation of the fisheries, still it remained nearly a dead letter, as the Government had appointed no magistrate near the bay who could be applied to in case of insubordination, thus obliging the manager, when such instances occurred, to go to Adelaide, a distance of fifty miles, for a warrant in the first place, and then to wait there until the offender was brought up, a period of many days. This caused many offences to be passed by which were very injurious to the success of the fishery."

The Act was rescinded in July 1842, together with an amendment of 26 June 1840. By this time the South Australian Company had begun to lose heart, and on 5 March 1841, George Fife Angas wrote: "The South Australian Company has five vessels employed in the black and sperm whaling trade besides small craft in the neighbourhood. We are withdrawing them from the sperm whale fishery. The coasting fishery, though the Company should withdraw from it, will fall into private hands." The private hands were Hack and Company, who took over the Company's half share of gear and boats and waited for the next winter season.

This year again brought foreign vessels to Encounter Bay; "It is well known upwards of thirty vessels are engaged between Kangaroo Island and Cape Lewin," according to Captain Hart.

Of all whaleboats in use, either at sea or in bay whaling, the Americans were reported to be among the fastest. During the Yankee invasion the locals were at a disadvantage in their heavier boats, built to withstand the pounding surf, and found it difficult to secure their fish. Foreign ships, anchored offshore or hove to a few miles out to sea with a masthead look-out, also had a good start when "white water" was called.

If the deep sea whalers had a dry season, their favourite trick was to skirt the shores of Tasmania, New Zealand, or South Australia during the winter months, and lay off from the largest bay whaling station they could find.

In an effort to beat the deep sea men at their own game, bay whalers often put out with enough supplies for two or three days. But if no whales were sighted, one night in a cramped whaleboat was usually enough, and the deep sea men lounging on the decks of their ships would cheer them back to land and await the arrival of the fish.

In 1842, the Encounter Bay fishery was listed under the name of Hogan and Hart, and a cargo of oil sent home by the *Daniel Wheeler* realized £34/15/- per ton. From then on reports were limited to occasional newspaper paragraphs until 1851, when the station under Messrs Boord, Bennett, and Johnson comprised sleeping berths, stables, boat-sheds, and workshops, about twelve buildings in all. Mr A. Long was named as superintendent.

By this time bay whaling was fading out in Tasmania and New Zealand, but it must have still been promising at Encounter Bay. In October 1854, the South Australian Government and the local District Council built a road skirting the Bluff, and a jetty for the use of whaling vessels. And in 1861 "shears" were erected to facilitate the cutting up process.

Compared with Tasmanian and New Zealand bay whaling stations, Encounter Bay was never really prosperous. The main causes of failure were inexperienced men and headsmen, the absence of really big schools of whales, the confined spaces which added fuel to rivalry, and the lack of law enforcement. Unfortunately, those who pioneered one of South Australia's first industries entered bay whaling at a time of great competition from deep sea vessels, and just when the yearly migration of whales was beginning to fade, it did rapidly from 1850 onwards.

Over the years, while the various station owners battled with profit and loss accounts, and whalers battled with each other for fish, two groups stood more or less on the sidelines at Encounter Bay; the Aborigines, and the permanent settlers headed by the Rev. R. W. Newland, who arrived on 30 June 1839 in the *Lord Hobart*.

Mr Newland's company settled in tents at Yelki, the Aboriginal name for the area adjacent to the Bluff, and immediately erected a rough chapel.

Very little has been recorded about the association between whalers and permanent settlers, apart from some mention in Simpson Newland's classic *Paving the Way*, but since they had little in common they probably pursued a policy of live and let live. However, whaling as a recognized industry gave the settlers a port, and a means of communication and transport between Encounter Bay and Port Adelaide. What little trade existed between the two places was carried on by vessels calling to pick up the produce of the whaling stations.

The Aborigines, on the other hand, took more than a passing interest in the whalers.

Before the arrival of the whalers, they had very often stood on the shore or climbed the Bluff to watch the whales diving and blowing out to sea. No doubt they entreated the spirit of the sea to send one ashore, and when a whale was stranded they descended upon it to gorge themselves.

Then suddenly the strange pale men came, and in their weird craft which they propelled through the sea with long sticks, actually tackled the monsters out in the bay and towed them back to shore.

But the Aborigines must have been amazed to see what happened when the first whale was caught and dragged ashore. Instead of falling on the carcass and relieving their hunger, which must have been great after such a tussle, the newcomers lit a fire, melted the delicious meat, and ran the liquid off into barrels.

The methods used at Encounter Bay in boiling down and flensing, however, were as primitive as at other bay whaling stations, and there was always plenty of flesh left over. The Aborigines were quick to learn, and often stood watch on the Bluff, and even cheered on the crews as they raced across the bay.

They had good reason to be excited, because the beaching of a whale was the prelude to a mighty feast. Dr Leigh describes the flensing process and the glee of the natives:

About sixteen men were appointed to the task, which is called 'flinching' him, and was thus performed: A man having alighted upon him, with a kind of sharp spade, began cutting lines across, and making a deep longitudinal incision in his back. A hook with pullies was then let down from the yardarm, and fastened to this portion, which was in the form of a belt. As the pulley wound it up, the men on the whale kept chopping away till a piece four feet wide and twenty feet long was suspended. This was then lowered into the sea and towed by boats to the shore, for the purpose of being boiled. The blubber, being on shore, and in the house where it is to be converted into oil, is taken up with forks, and chopped into slices, thrown into the coppers, and the oil ladled off. The scratchings, or refuse after boiling, serve for fuel. The entrails, &c. of the whale float to the shore, and by the action of an Australian sun, send forth the most abominable effluvia imaginable, attracting all kinds of carrion birds, not forgetting blackee, who is now in the height of his enjoyment.

I met many scores of natives, marching in high glee to

the scene of action; for no sooner is a whale killed, than the affair literally "gets wind," and the gentry from the Black River, and the interior descend in flocks. Then was blackee in his element. As I strolled along the beach, what scenes did I behold! Some chattering and gabbling as they were striding to the feast; others, sitting round a fire, with an oven made of stones, upon which was the blubber, while they intently watched the progress of the cooking, and inhaled the delightful steam that issued from it. Some, who had eaten their fill, lay like lifeless carcasses, round as a barrel, upon the sand; and numbers were carrying large square pieces to their "lubras" upon their heads, putting out their tongues, ever and anon, as the delectable "gravy" melted from the blubber. The whale is to them as good raw, for their impatience never permits it to be more than warm through, before they draw it from off the stone fire and devour it. Every one screws off a piece, and crams it down his throat. They never cease eating till unable to swallow another morsel, when those who are able, arise, and bury the remains, which are disinterred when appetite returns.

No alderman of forty years standing ever devoured green (turtle) fat with more gusto than did those savage disciples of Epicurus the delicate morsels of the whale.

Encounter Bay natives also figured in the last attempt to revive the industry in 1870, when Mr B. Ranford manned his boats with Aboriginal crews, but from one account, when nearing their prey, they complained that their arms "were big one tired."

Mr Ranford must have been a courageous man. By this time bay whaling had lapsed in Tasmania and New Zealand, but whales must have still been in the Encounter Bay area in sufficient numbers for him to lay out his capital. *The Register*, of 7 October 1872, gives this glimpse of the last whale caught at the bay:

The crew from Mr B. Ranford's fishery went to sea a short time ago looking for whales, and after being out for a considerable time without finding one returned to Granite

The head of a sperm whale, hoisted onto the deck of a whaling ship (*Mitchell Library*)

The tryworks on a whaling ship, with oil flowing from melted blubber in the cauldron. The chimney could be turned so that smoke would drift downwind (*Mitchell Library*)

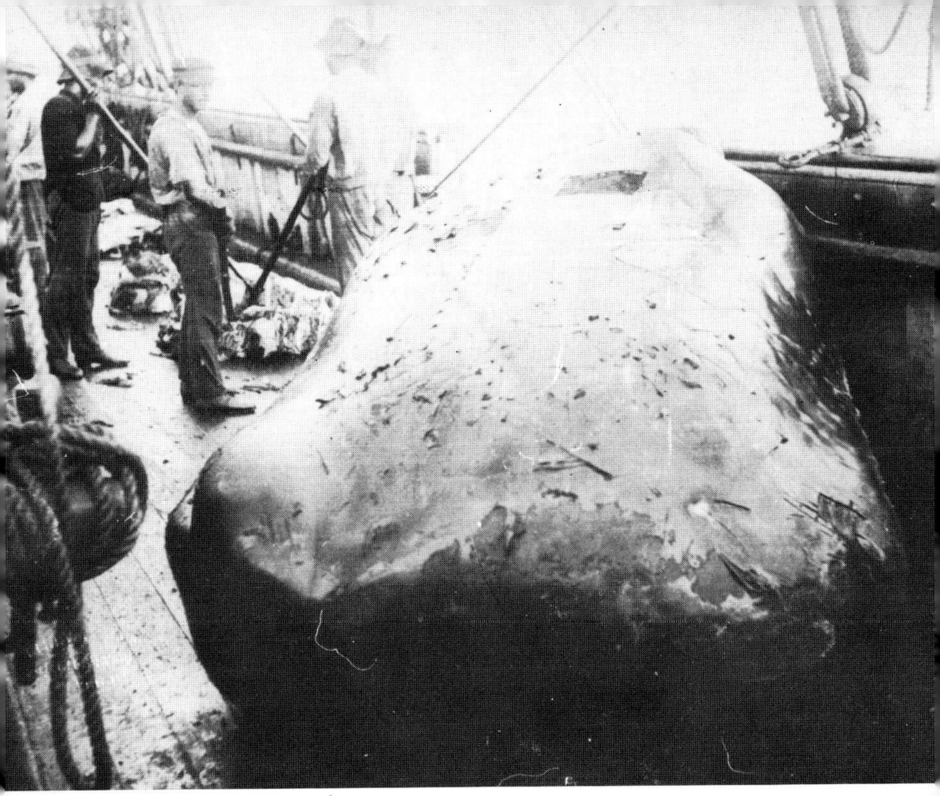

Island to rest. Whilst they were there a whale came into
Victor Harbour and showed itself above water near to a
barge that was loading the brig *Centaur* at the moorings.
The signal was quickly given and the crew as rapidly were
in the boat. Having pushed off from the Island they had
scarcely made the third stroke with the oars when they
reached the unexpected monster. John McCarthy soon
fastened to it; then A. Ewen the chief headsman, an
experienced whaler, although he had not struck a whale for
many years, was up to the mark and with superior skill
his lance dispatched the monster of the deep, although it
took the boat about two miles out to sea. A second boat
from the station started to assist and the creature was taken
to Rosetta Jetty about dusk. Next morning early horsemen
and vehicles were going from all directions to the fishery,
and throughout the day there was a continual stream of
visitors. No doubt Mr Ranford will be encouraged to
persevere for another year and we wish him success in his
spirited enterprise, but it is to be hoped that he will get
another whale or two before the season closes.

The good wishes were not consummated, and this is the
last recorded report of whaling in the area. Although whales
occasionally appeared in the bay in later years, there was no
one on the Bluff to raise the alarm.

The *Kyo Maru No. 15*, of 738 gross tons. She is one of the fleet of whale-catchers
attached to the whale-factory ship *Kyokuyo Maru* (*Kyokuto Hogei Co.*)

The Japanese whale-factory ship *Kyokuyo Maru*, gross tonnage 16,433. Note the
square access port in the stern, through which whales are heaved up on deck
(*Kyokuto Hogei Co.*)

7

Young bachelors of capital

"THERE IS UNFORTUNATELY IN ALL OF US A GERM OF selfishness which, when unchecked by public opinion or political opposition, is apt to grow into injustice and tyranny. In private life the squatters were excellent, generous hospitable men; but one large proportion consisted of old colonists accustomed to convict servants, who cost nothing beyond their board and lodging, and another of young bachelors of capital, who arrived in the colony to make a fortune, intent on returning to the old country as soon as it was made. The one despised, and the other were indifferent to the opinions of the working classes. Both dreamed of naturalizing in Australia the miserable wages of the southern counties of England and the Highland counties of Scotland." So wrote Samuel Sidney in his book *The Three Colonies of Australia*, published in 1853, and his condemnation could have included Benjamin Boyd. Born at Merton Hall, Wigtonshire, Scotland, about 1796, Ben Boyd was a bachelor, although not very young. He was forty-six when he arrived in Australia, but he certainly had capital.

While Samuel Sidney had an axe to grind with the landed

70

gentry who exploited the labour market of the day, Ben Boyd had an ambition that wrote one of the most colourful chapters in the history of Australian whaling. A tall, bearded man, with a rather thin pointed face and high forehead, he had, as the artist and diarist Georgiana McRae recorded in 1842, "the sanguine temperament, exuberant vitality, and daring enterprise of the typical adventurer."

As a business man and banker, his eye turned to the eastern coast of Australia as early as 1839. The area was reputed to offer attractive opportunities for men of drive and capital, but his first two letters to the British Government, asking for grants of land, scarcely raised an eyelid. But he was undeterred, and before he left London he floated two large companies; the Royal Bank of Australia, and the Australian Wool Company. The bank alone had a capital of over one million pounds, partly secured from small Scottish investors. In September 1840, he despatched the paddle steamer *Seahorse* from Gravesend for Sydney, followed by the *Juno* and the *Cornubia*, all carrying a quantity of wines and spirits. Boyd himself travelled in his private yacht *Wanderer*, first to Melbourne and then on to Sydney.

The *Wanderer* and her owner were prepared for anything. A sleek topsail schooner of eighty-four tons, she had a picked crew and carried thirteen guns. As a unit of the Royal Yacht Squadron, she was free to enter and leave any British port without restraint or control. The *Sydney Herald* described her arrival in Sydney on 18 July 1842 as follows: "Arrived from Port Phillip, having left 12th instant, Royal Yacht Squadron schooner *Wanderer*, 84 tons; B. Boyd, Commander. Passengers: James Boyd Esquire, Messrs Brierly, Bogue, Goddart, Downes and Prathent. On coming to anchor in the Cove the *Velocity* schooner belonging to Mr Boyd, fired a salute and the neighbouring heights were crowded with spectators to witness her arrival. The *Wanderer* is armed to the teeth, and is fitted up in the most splendid manner; in fact she fully answers the description which has been lavished on her by the Port Phillip Journals."

While the onlookers, envious of the life of the idle rich who could afford to sail the world in luxury, went back to their homes, Ben Boyd began putting his plans into operation. First the *Seahorse* went into service between Sydney, Port Phillip, and Launceston, and then he bought large tracts of land on the Monaro Plateau and in the Riverina district.

By 1844 he was one of the biggest squatters in the country, with eighteen properties in two states. His holdings in New South Wales alone totalled one and three quarter million acres. As an outlet and port he chose Twofold Bay, 240 miles south west of Sydney, named by George Bass in 1798.

Like many bays on the Australian coast, it had been used by sailing ships as a haven during storms long before it was officially named. French, American, and English sperm whalers had also set up their tryworks in the bay at different times, to catch up with their "trying out," and on the 15 August 1828, it figured as a news item in the *Sydney Gazette*:

> Mr Thos. Raine of Bligh Street is an enterprising colonist and one of the most indefatigable merchants Australia can boast. About three months ago it occurred to Mr Raine that something might be done at the hitherto untried whaling ground at Twofold Bay. An experiment on a small scale was started by a Mr John Irvine—Australian by birth and a ship builder by trade. The schooner *Darling* was sent there and returned last Tuesday after only three months, with sixteen tons of oil. But this is not the only news we have. We are informed that the whales flock into the bay in shoals owing to which there are now between eighty and ninety tons of oil on the beach and ready to be despatched to Sydney the instant the proper vessel is despatched for it.
>
> Such an adventure as this we should presume will lead more to open the eyes of the public as to the resources of Australia than all the newspaper remarks which might be put forward for a century to come.

The writer was correct. Four years later, a Tasmanian merchant, Peter Imlay, on his way from Sydney to Launceston and anxious to expand his activities, decided to settle at Twofold Bay. He was later joined by his brothers, Dr

Alexander and Dr George Imlay (also young bachelors of capital) who laid plans for a scheme which eventually brought them into direct competition with Ben Boyd. Like Boyd, they secured large tracts of land, with a view to exporting sheep and cattle to Tasmania as well as setting up a bay whaling station.

When Boyd arrived, he found a prosperous whaling station on the beach and a small township on the north side of the bay. Ignoring them, he chose the southern shoreline as the site of his township.

But his was the grand scheme; no pug wall shanties built by amateurs, or ships lying offshore to be loaded by small boats; he had the capital and he used it lavishly. In the *Harlequin*, he brought down from Sydney eight carpenters, four stonemasons and bricklayers, one plasterer, one smith, two stockmen, and fourteen labourers and their families.

This was the vanguard. Many more joined him later, and in 1848 William Henry Wells in his *Gazetteer of the Australian Colonies* gives this description of Twofold Bay:

Boyd Town, although but lately founded, is already a flourishing sea port, enjoying a commerce of considerable importance, and being the key to the extensive Maneroo country (whence an excellent road has been constructed) it is the chief port of outlet for the south eastern districts of New South Wales. The first object which strikes the eye when approaching Twofold Bay from the north, after passing the Dromedary, is the light-house now erecting on the south head of East Boyd, and which is nearly completed. It is a tower composed of white Sydney sandstone, in blocks averaging half a ton each in weight. Its height, when finished, will be seventy-six feet, its diameter twenty-two, and it is not an exaggeration to state that it is the finest building of the kind in Australia. Its use as a beacon in the day time will be obvious, as by reason of the height of the headland it can be seen at a distance of twenty-five miles. At East Boyd is the large whaling establishment of Mr Boyd; whence nine sperm whalers now sail; and inasmuch as Great Britain and all her colonies have only fifty-nine vessels engaged in this important trade, which, in the Pacific alone, employes nearly 700 American whalers, the most correct idea of the value of the depots at Twofold Bay, is thus given. At Boyd Town there is a convenient jetty, 300 feet long; and as vessels seeking the port to refit have the advantage of a heaving-down hulk, and every necessary mechanical assistance—abundance of water, and every description of provision and vegetables—both Boyd and East Boyd are favourite resorts for shipping. The laying out of Boyd Town is in good taste. A handsome Gothic Church, the spire of which is visible twenty miles at sea—ranges of commodious stores, some 120 feet in length—well built brick houses, and neat verandah cottages—a splendid hotel in the Elizabethan style, large salting and boiling-down houses, mark the rapid advance of this young and hitherto almost unknown port of the Pacific.

And so Ben Boyd, having provided facilities for visiting whalers, a deep sea port as an outlet for his pastoralist activities, and having established his own deep sea whaling fleet as well as his own bay whaling stations, began earning such compliments as appeared in *The Sydney Record* of 24 February 1844: "Perhaps there is no individual who has done so much for the colony and in so short a time as Mr Benjamin Boyd. We understand that the *William, Fame,* and *Juno,* lately purchased by him, are now undergoing thorough repairs and are fitting up as whalers. These ships will shortly be ready for sea. We hope success may attend their voyages, for if there is one gentleman more than another who deserves to be amply remunerated for his enterprising speculations, that gentleman is Mr Boyd. If we had a few more such men in the colony it would soon go ahead."

Ben Boyd, endowed with capital and a flair for organiza-tion, was one of the first to exploit the country's natural assets of good grazing land, the wealth of the seas, and natural safe ports, which he welded into one grand scheme. Wherever he went he was treated with awe and respect;

he entertained lavishly aboard his yacht and on shore at Twofold Bay. Australia had never seen his like before, but like all empire builders he made his enemies. Samuel Sidney was perhaps the most vehement, saying:

Mr Boyd arrived with the express purpose of making investments at the time (1841) that the colony was in a general state of insolvency, or, as he expressed it 'in a jam'. A yacht of the Royal Squadron, an apparently unlimited capital, an imposing personal appearance, fluent oratory, aristocratic connexions, and a fair share of commercial acuteness, acquired on the Stock Exchange, at once and deservedly placed him at the head of the squatocracy. His aim was the possession of a million sheep. He was the chief of the hundred thousand sheepmen, with whom he combined to obtain fixity of tenure for their sheep pastures, to put down small settlers, and to reduce wages.

At the period we are describing, from 1841 to 1844, the colonial labour market presented the most curious contradictions. The bounty agents were pouring in a crowd of most unsuitable persons, who, once landed, were soon left to shift for themselves. Among the merchants of the town of Sydney distress prevailed in consequence of the cessation of building and other works, the wages of mechanics were depressed to a rate before unknown, and newly-arrived immigrants were astonished at the low rate of pay for town labour, so different to the flaming representations of the crimps by whom they had been collected. But in the country districts, and especially in the bush, where sheep and cattle were breeding, while their proprietors were going through the insolvent process, wages were maintained; and the anomaly was presented of large bodies of men being employed at the expense of government, at high wages, at public works, on a sham labour test, while flocks were wanting shepherds in the interior. Several causes supported this anomaly: First, there was no government machinery for distributing newly-arrived emigrants; secondly, the preference of the squatters for single men left families on the hands of the government; thirdly, the Squatters' Club was not sorry to see the government embarrassed by the presence of a large body of unemployed labourers in Sydney; fourthly, the dishonest conduct of certain masters in withholding, or unfairly deducting wages promised, had given the bush a bad name; fifthly, many of the emigrants were of a class who, having left parish aid behind, liked to keep close to government rations and wages. All were engaged, as far as their short-sighted views would permit, in killing the golden goose of colonization.

Mr Boyd's evidence before the immigration committee of 1843 affords, when read with the notes we can supply, a fair specimen of the haughty, gentlemanly, selfish class he represented. He had then been eighteen months in the colony, and was employing two hundred shepherds and stockmen, besides artificers. He was building a town and port at Twofold Bay; had two steam-boats, and a schooner-yacht, the *Wanderer*. He had devised a wild scheme of saving labour, by putting three thousand sheep, instead of eight hundred, under the charge of one shepherd, on horseback.

Mr Boyd despaired of the prosperity of the colony "unless the wages of a shepherd could be brought to £10 a year, or about 3s. 10d. a week, with meat and flour, without tea and sugar." The two last had been previously universally allowed; but he expressed his intention of doing away with them, "being of very questionable utility and necessity, although such is the waste and extravagance here that 8 lbs of tea and 90 lbs of sugar are consumed per head." He states, further, that he "had no difficulty in engaging shepherds at £10 with these rations, but much difficulty in getting men engaged at these low wages forwarded to stations, as they were generally picked up on the road. Any money advanced towards travelling expenses was usually spent in public-houses;" and it is his decided opinion that "more than £10 a year only does harm to shepherds, by sending them to public-houses."

Mr Boyd also mentioned how he had kindly given a free

passage to Twofold Bay, some 600 miles from Sydney, to one hundred labourers out of employ. He did not mention that, on their arriving there, those who refused to accept £10 wages were refused a passage back for less than £5; and that, while a few strong men walked back over the mountains, those who remained created such a feeling in the country that Mr Boyd could not venture to visit his stations until the time of the year when the police magistrate, with a guard of policemen, took his annual round.

Fortunately all squatters were not like the Boyd clan, and the productiveness of the land defeated the combination. Had it been otherwise, a very few years would have produced a servile war of men against masters. From the Boyd clan proceeded stories founded on fact, and dressed to suit a purpose, about allotments of land sold for quarts of rum, champagne drunk in buckets by shearers and shepherds, who insisted on having "pickle with their (measly?) pork."

Whether or not Ben Boyd "despaired of the prosperity of the colony" in his wild schemes, and exploited his employees, or merely took advantage of existing conditions to lay the foundations for a future empire which would eventually benefit employer and employees alike, was a favourite drawing room topic of the time, and one which was made more controversial by his untimely demise.

Before dealing with his fate, however, there are several interesting points about the physical side of whaling at Twofold Bay which deserve mention. The bay itself is really two bays in one, made up of two headlands forming the first fold and two others, Lookout Point on the northern shore and Honeysuckle Point on the southern, making the second. All were manned as look-outs, and other look-outs, posted at the masthead of ships lying in the bay, made it almost impossible for a spout to go unnoticed. This, plus the fact that whales were never quite as plentiful in the area as in other parts of Australia, set the scene for some of the most violent struggles in the history of Australian shore-based whaling. At times, as many as twenty boats from shore stations and ships surrounded the one whale. The men were on the same small lays as prevailed in other stations, but amid the shouts of the chase there was heard the jabberings of Australian natives. Both stations used them in their crews, and they proved themselves in their ability to row strongly in short bursts; ideal for whaling in the confined spaces. Their vision was also stronger than their white partners, and they had no need for telescopes when manning the look-outs. During the whaling season they lived the normal life of a whaler but at the end of the season were content to return to the bush leaving the white whalers to argue and squabble over past events and highlights of the season.

Ben Boyd was not the only one to take advantage of cheap labour. The Imlays paid their Aboriginal crews with provisions and tobacco; much cheaper in the long run than the 'lays' of the white men. Mr Wells wrote that: "The advance in civilization made by the Aborigines of Twofold Bay, since the formation of the whaling establishments, is not the least interesting feature in a sketch of the progress of the place. They are an active and intelligent race, and in their useful labours, in boating, and in various arduous employments on board the whalers, they certainly contradict the hasty conclusions which so many superficial writers have drawn in reference to the degrading tendency of the faculties of the natives of New South Wales."

One wonders if Mr Wells did not turn a blind eye to the scene of Twofold Bay natives gorging themselves to stupidity on whale offal when he wrote such a glowing report of their behaviour. The only handouts they could expect around stockmen's camps and homesteads were occasional flour and tobacco, but near bay whaling stations the handouts or leftovers from the kill were to them the most delicious meal imaginable, which they ate with reckless abandon. Little wonder that they were always interested bystanders, if not active participants, and played the part of unofficial look-outs whenever possible.

At Twofold Bay their eyes wandered from the whaleboats ready at the base of the cliffs out across the expanse of the bay. While the natives watched, the white whalers attended to their boats, reputed to be among the fastest on the Australian coast.

These were of the usual construction of huon pine, and most pulled five oars, but it is on record that some used seven and at least one of them shipped nine oars. From all descriptions it is evident that the whalers took more than the usual pride in their boats, and a good boat, proven in a season, often changed hands at keen prices. From all accounts the whalers in general at Twofold Bay were more skilled and more reliable than their counterparts at Encounter Bay, perhaps because they were under the immediate critical eye of their overseers. On the coast of Tasmania and South Australia they were often isolated, a law unto themselves, but here they were part of organizations run on strict business lines.

More often than not they had an audience of interested spectators leaning over the railings of ships in the bay. Oswald Brierly, Diarist and Artist, who came out with Ben Boyd on the *Wanderer* and was put in charge of his station at the bay, has left us this record: "After the whale had been fastened, it turned towards the boat, and made a dive but came to the surface exactly underneath us. I shall never forget the new and sickening sensation of feeling the whole boat suddenly lifted out of the water by the rising back of the whale and the sliding, gliding, hopelessly slipping motion of the boat as it shot down the back of the monster. The powerlessness of the situation flashed through the mind—the oars were of no use, we were literally out of our element, but only for a moment and then as we touched water again down came the great tail of the enraged monster cutting our boat in twain with one swishing blow as a hatchet would chop a pat of butter. In a moment the whole party was plunged into a boiling eddy of foam and water made by the whale as it plunged into deep water and left us struggling for dear life and handicapped by a thick coat and heavy sea boots." Such spectacles must have been an added attraction for many who took passage up the eastern coast of Australia on ships calling at Twofold Bay and one can imagine what the modern advertising agencies handling shipping publicity would make of it today.

No doubt the whalers themselves responded to their audience and acknowledged the cheers as they towed their carcasses back to the beach for trying out.

Apart from this they also had the opportunities of more civilized living when compared with other more isolated shore-based stations. Bruised and aching limbs could be soothed with a draught at the Seahorse Inn, which perhaps provided female companionship of the right type for men who daily risked their lives and limbs on the bay.

During his years in Australia, Ben Boyd, no armchair administrator (he took part in at least one whale hunt out on the bay), spent his time travelling between Twofold Bay, Sydney, and his pastoral properties, and it is not difficult to imagine the problems of running a scattered business in those days. His whaling fleet used Twofold Bay and Mosman's Bay, which in the 1840s was the main ship repairing depot in New South Wales, while his coastal vessels brought wool from Twofold Bay to Neutral Bay (where he had built a wool washing dam) to be prepared for shipment to countries overseas.

Although he delegated authority, the enormous amount of clerical work and the delays in receiving replies to important despatches must have been irksome to say the least. His overseas markets, like his latest market price reports and his shareholders, were months away by sail. The latter may have been an advantage, but it certainly would not lessen his responsibilities in manipulating their investments, or ease the difficult task of choosing men to take charge of individual stations or projects at Twofold Bay.

Then, too, he had to contend with the ever present risks of disasters at sea.

Since the *Seahorse* had arrived in Australia, she had been making regular trips to Port Phillip and Hobart Town, and on 5 June 1843, the following entry is recorded in her log: "At George Town Cove, got on shore when the water was ten feet forward under her, and six feet aft; the ship heeled over at low water and strained very much; her butts above water appeared injured and strained. At flood she righted and got off." After reaching Sydney she was laid up, and Boyd claimed £25,000 for total loss from the Royal Assurance Company and other companies which had insured his ship. The companies refused, and Boyd brought an action against them for £5,000. This led to a long and costly lawsuit in London, during which a special committee was sent out to Australia to examine witnesses and carry out closer investigations.

Even in those days it appeared there were welcome fringe benefits in big business, and one must come to the conclusion that Ben Boyd, a born entertainer, could not resist the temptation to throw a party. Wine flowed, and Mr Boyd was in the chair during a dinner on board the *Seahorse* for the inspection of the ship. When this came out in evidence in London it caused some amusement and no doubt helped to weaken his case. His claim was not allowed.

To the people of New South Wales, Boyd's arrival and initiative, the building of his port and town, and the establishment of his whaling fleet in such a short time, was little less than a miracle. But those with their fingers on the pulse of the economy shook their heads; he had over-extended himself. The clouds of a depression were already building up when he arrived, and by 1846 stations were being sold for a song. By 1848, his shareholders had become dissatisfied with his administration, and demanded his replacement.

He was succeeded by his cousin, William Sprott Boyd, who inherited a financial tangle which would make a modern businessman shudder. During the setting up of his empire, Ben Boyd had expended vast sums of money and paid wages to hundreds of workmen. To offset the drain on ready cash he issued private notes to the value of five shillings up to one pound, which could be cashed at stated places. Holders of these notes usually bought goods at his stores. The auditing of this system alone would present frustrating problems, while the complexities of the association between the Royal Bank of Australia, the Australian Wool Company, and his London agents responsible for the purchase, exporting and eventual sale of the produce from his pastoral and whaling activities, left Sprott Boyd with no alternative but to wind up the business to protect the woeful shareholders from further losses.

By 1849, Ben Boyd's operations at Twofold Bay had ceased. Many of the buildings had never been completed. His ships, shore whaling equipment and boats passed into other hands, and it is ironical that his lighthouse, surely built by Boyd as a beacon to blazon his achievements, never showed a light.

Boyd, however, was not the only one to fail at Twofold Bay. His competitors, George, Peter, and Alexander Imlay, who had taken up large tracts of land at Bega, were in financial difficulties by the end of 1844. In December 1847 George, who had contracted an incurable disease, shot himself on top of a mountain near Bega, now known as Dr George Mountain.

They and Ben Boyd, all bachelors of capital when they arrived in Australia, were men of vision who tried to combine whaling and pastoral interests. Had they concentrated on one or the other they may have succeeded, but they spread their interests too widely in a competitively depressed economy.

When compared with the glowing accounts of his arrival in Sydney, there is something very final in the terse announcement of Ben Boyd's departure which appeared in the shipping columns of 26 October 1849: "*Wanderer*, R.Y.S., Mr B. Boyd owner."

His thoughts, as he sailed out through Sydney Heads, are unknown. Perhaps he was cursing his luck, the thin thread separating business success and failure; perhaps bemoaning his stupidity for passing so much authority to others; or perhaps remembering with pride his role in the social life of the colony. Or maybe he was simply looking ahead, for fresh worlds to conquer.

Ben Boyd set a course for California. A year later, the *Wanderer* lay off the island of San Cristoval. The commander went ashore, gunfire was heard by her crew, and signs of a struggle were found later by searchers, but Ben Boyd was never seen again.

On 13 November 1851, the *Wanderer* returned to Australian

waters, a battered hulk after weathering gales. In attempting
to enter Port Macquarie she ran aground and became a
complete wreck, and the last link with Ben Boyd, the young
bachelor of capital who had hoped to build a financial
empire in which whaling would play an important part,
was broken.

8

Favourably disposed towards the killers

"Of this whale little is precisely known to the Nantucketer, and nothing at all to the professed naturalist. From what I have seen of him at a distance. I should say that he was about the bigness of a grampus. He is very savage—a sort of Feegee fish. He sometimes takes the great Folio whales by the lip, and hangs there like a leech, till the mighty brute is worried to death. The Killer is never hunted. I never heard what sort of oil he has. Exception might be taken to the name bestowed upon this whale, on the ground of its indistinction. For we are all killers, on land and on sea; Bonapartes and sharks included."

From this description by Herman Melville, it seems that whalers steered clear of the killer whale, or Orca, and for this reason the naturalists who occasionally accompanied sperm whalers on their voyages had little opportunity to examine the killers at close quarters. One can sympathize with these dedicated men. Their only chance of furthering their knowledge was to board a whaler for two or three years, and while they usually rated a cabin to themselves, their interest in nature alone separated them from the men whose

A harpoon-gun, loaded with head containing an explosive charge and with barbs which fly open inside the whale. The design is basically the same as that of the gun invented by Sven Foyn, which revolutionised the whaling industry and robbed the whales of their "sporting chance." The other two photographs show the gun in action (*Nor' West Whaling Co. and Kyokuto Hogei Co.*)

main concern was grease, and they were forced to endure great hardships for the cause of science.

And yet such men as Thomas Beale, who sailed as surgeon in the *Sarah and Elizabeth* in 1838, were able to make observations on marine life which have stood the test of time. His book *The Natural History of the Sperm Whale* is an example of his keen observation and descriptive powers, but like others of its time it says little about the Orca.

But Herman Melville himself was in many ways a naturalist at heart, and two of his observations on killer whales proved to be correct. They are certainly savage, and whenever possible they attack whales; next to man, the killer is the whale's worst enemy.

Like most fearsome creatures of the sea, the killer has been surrounded by exaggerations and fables for centuries. One story, that they slit open their victims' bellies before devouring them, survived almost as long as whaling from sailing ships. Another, that they swallow their victims whole, is founded on fact; whole seals, penguins, and porpoises have been found in their stomachs. According to the naturalist E. J. Slijper, a killer dissected in the Bering Sea contained two full grown seals.

This may seem hard to believe, but the killers are certainly equipped for such carnage. Growing to a length of thirty feet, with fins which cut the water like a knife, they are equipped with ten to fourteen conical teeth on each side of the upper and lower jaw. Like wolves, they have stomachs which can accommodate large quantities of unchewed food. In 1956, the Icelandic Government paid the United States Navy almost £100,000 to bombard schools of killer whales with depth charges, to save their southern coast fishing grounds. They attack anything that moves and seem to have little fear of the presence of man or his equipment, and old time whalers could do little if killers decided to attack a carcass brought alongside for trying out.

When they attack whales at sea, they concentrate on certain parts of the body. As Melville pointed out, they confine their attacks to the tongue and the lips, tearing off great chunks until their victims die of loss of blood and

Blue whales on the deck of a Japanese whale-factory ship (*Taito Fishery, Tokyo*)

These Japanese seamen are flensing a whale; cutting into the blubber so that long strips can be pulled off by a winch (*Taito Fishery, Tokyo*)

exhaustion. Like most marauders they hunt in packs, and whales, despite their size, have little defence against them. Their only hope is to leap out of the water, and endeavour to scatter their opponents by threshing around with their tail flukes. Adult sperm whales, however, seem to escape their attacks, and most sperms, particularly bulls, are free of the scars of killer battles. But many old whalers have left eyewitness accounts of sperm calves being killed and eaten within a few minutes of a pack attack. Whalebone whales, which included the right whales of Australian waters, were even more defenceless when they were in the shallow inlets and bays of their breeding grounds.

Most old whalers' tales emphasize the killer's intelligence, and some his playfulness. They tell of them hanging on to whale lines and being towed through the water; of waiting around ships during "trying out" for the tongues to be thrown overboard; and of their having homing instincts, returning to the same bays year after year, a few days before or after the arrival of right whales. But nowhere is there any record of their co-operating with whalers for the destruction of their prey—except at Twofold Bay on the eastern coast of Australia.

There is ample proof that pack hunters on land have clearly defined methods of strategy. They drive their prey into a confined space, then divide to cut off retreat, and the rear guard shares in the spoils after the kill. And it is reasonable to believe that killer whales do the same; they certainly surround their prey, and afterwards the whole pack shares in the meal. But to suggest that the killer whales of Twofold Bay actually drove whales into the bay, held them in captivity, and if for some reason the whaleboats were not at the ready, sent scouts close in shore to rouse the whalers from their sleep or drunken stupor, seems unbelievable. But according to documented evidence, this is just what happened.

After the eclipse of Ben Boyd, whaling at Twofold Bay was carried on by the Davidson family who over the years learnt to recognize many of the killers by sight. There was Old Tom, who had a peculiar mark on his fin, and Hookey with a bend in his dorsal fin, and Humpy, and Stranger; a few of the thirty or more killers who came every year to the Bay. When questioned over the years by various writers of magazine articles and authors, among them W. J. Dakin who wrote *Whalemen Adventurers*, the Davidsons and their men treated their alliance with the killer pack quite casually.

Old Tom, they said, had an annoying habit of hanging on to whale-lines, and enjoyed being towed through the water. But the most amazing stories deal with the strategy used by the packs in general. When the whales arrived they actually divided themselves into three packs; the first stayed close inshore, the second at a middle distance, while the sentries posted themselves well out in the bay. The sentries allowed the schools to pass to be driven into the bay by the middle group. This in itself saved the whalers hours of backbreaking rowing, and allowed them to hunt in the relatively calm waters of the bay instead of in the open sea. Then, according to accounts, four killers separated from the pack, and while two of them swam beneath a whale to prevent it from diving, the other two worried it on the surface until it was exhausted.

But if a whale came into the bay and no boats were out, one killer would detach himself from the pack, and swim inshore to rouse the whalers by flapping his tail on the surface of the water. When, as often happened, boats were capsized or smashed and whalers were thrown into the water, the killers would hover protectively around the men until they were picked up. No record has ever been found or no story ever told of a Twofold Bay whaler being attacked by a killer whale.

If any of the above doings of killers had been included in old whaling books or accounts of voyages, or even log books, they would have been viewed with great suspicion and dismissed as the ravings of drunken men or exaggerations from which myths and legends grow. W. J. Dakin appreciated this and questioned men, some of them 300 miles apart, and in every case the stories followed the same pattern. But still he was not satisfied and was the first person to closely examine the notes and diaries of Sir Oswald Brierly written in 1843, who was closely associated with Ben Boyd and the Imlay Brothers period. These diaries are now in the Mitchell

Library and as Dakin points out, do nothing to disprove the doings of the killers in the time of the Davidsons. One passage of the diary reads: "They (the killers) attack the whales in packs and seem to enter keenly into the sport, plunging about the boat and generally preventing the whale from escaping by confusing and meeting him at every turn, The whalemen of Twofold Bay are very favourably disposed towards the killers and regard it as a good sign when they see a whale 'hove to' by these animals because they regard it as easy prey when assisted by their allies, the killers." Dakin further points out that the diaries were taken to England in 1848 and none of the Twofold Bay whalers questioned in later years saw them. He believes that their stories of the doings of the killers were true in every respect.

And what of the killers today? Many of them disappeared when the right whales stopped coming to the bay, but one, Old Tom, continued to put in an appearance. He was easily recognized by a white band encircling his body just behind his long dorsal fin. In September, 1930, he was seen floating on the surface, and the next day his carcass was washed up on shore.

The people of Twofold Bay had two choices; they could allow the carcass to rot and put up with the smell, or have it towed out to sea. They chose a third course when Mr Logan, one of Old Tom's friends, suggested that his bones be preserved. Today, Old Tom is suitably housed in the main street of Eden, the local town. This in itself is proof of the high regard which the locals had for the killer packs of Twofold Bay, and even today they dismiss tourists who remain dubious of the killer tales with a shrug of the shoulders. Perhaps, when the skeleton of Old Tom has crumbled to dust, they will be dismissed as fables and legends, but today they are believed.

W. J. Dakin also discovered another interesting passage in Oswald Brierly's diaries which throws some light on the attitude of Australian natives to the sea creatures that inhabited their coastal waters.

The myths and legends of inland tribes, woven as they are around animals, birds, and reptiles, which figured intimately in their daily search for food and survival, are charming and in some cases almost believable; the kookaburra was chosen by the sky spirit to herald the coming of the sun and to waken those who may have overslept, and the soft light of the moon was installed in the heavens to save people stumbling around in the dark.

No wonder, then, that the killer whales who must have driven many a whale ashore at Twofold Bay before the coming of the white man, thus providing a royal feast for the local natives, should also have their place in Aboriginal legends.

Brierly's passage reads: "The natives of Twofold Bay regard the killers as incarnate spirits of their own departed ancestors, and in this belief they go so far as to particularize and identify certain individual killers."

And so it was that even the natives of Twofold Bay were favourably disposed towards the killers.

9

To seek it at a distance

Ship ALBION, *at sea, 5 October, 1803. I am happy to inform His Excellency that I have bin a Vandaman Land, and landid the stock and stowers on acc't the Government. We had 12 days' passige, with 3 days that we lay in Oyster Bay, with light airs from Sd. We obtained 3 sperm whales in sight of Oyster Bay. His Excellency mention'd of landing the stock at Ralph's Bay, but Mr Bowen and myself thought it most prudent to run up within 3 miles of Resdon Cove, and with lashing 2 of my whaleboats side to side, we got them on shore very well. . . . I hope Mrs King and the little gairl is well. Governor Bowen sends Mrs King a p'r black swans, which he bages her acceptance.*

I have, &c.
E. Bunker.

P.S. I thank your Excellency to let me have 150 Ackeres of land more at the Hawkesbery River, as I have only tackin 50 at Vandaman Land.—E.B.
I send your Excellency a p'r of swans, which I bage his exceptance.

Captain Ebor Bunker, who took the first convicts and

stores to the Derwent and who later became a well known Sydney whaler, also had ambitions towards the land, as the above letter to Governor King proves. Whether a pair of black swans changed hands with every land deal is not known, but by the time he died in 1836, he owned some 1,500 acres. He was not the only whaling captain to use his rather unique position in the colony to advantage. As mere seamen they would have been unacceptable and quite out of place at social functions at Government House, but as captains, many of them representing overseas companies and capital, they were often called in for consultations by colonial governors.

Apart from this, merchants clamoured for news of market prices and business trends in far off countries; whaling captains, sailing from London for Port Jackson and Hobart Town, could deliver a letter in under five months; Captain Bunker made a voyage in three months fifteen days. They were an important link with the business hub of the British Empire, and the sight of their sails from Sydney Heads stirred the commercial world of the colony. They were also the bearers of international news. Minor crises could erupt and be settled or continue over long periods, while colonial governors and merchants entertained potential enemies within their harbours.

While their ships were being refitted and stocked for the whaling voyage, captains enjoyed their brief hour of glory, at least until another ship and another captain pushed them into the background. But during their stay they were in a good position to see something of the colony and measure its progress since their last visit. Their impressions induced others to seek their fortunes in Australia. Ebor Bunker himself, an American, must have been suitably impressed, not only with the possibilities of whaling but grazing as well. Ben Boyd felt the same way, and the Imlay Brothers at Twofold Bay also combined pastoral activities with whaling, as did the Henty Brothers in Victoria, The South Australian Company in South Australia, and other companies in New South Wales. This spreading of capital did not always bring the desired results for the companies concerned as we have seen, but it did help to create a variety of opportunities for colonists.

However, in the early years of whaling, captains had no qualms about competing for labour with landowners as Governor King points out in a letter to the authorities in London in 1805. "Another cause for the want of agricultural labourers is the number of men who are employed by individuals in the seal and oil fishery."

Within half a century this trend was reversed, as men left the sea to "dig the land" but in Governor King's time all eyes were on the sea and the harvest it could bring. But despite this, the New South Wales colonists were rather slow off the mark as far as deep sea whaling was concerned. This was understandable; unlike bay whaling, deep sea whaling called for considerable funds for the purchase of ships, boats, and equipment. If they were to be built in the colony it needed shipwrights and tradesmen of many kinds, as well as local funds. These were not always available in the early days, and for the first thirty years the British and Americans enjoyed almost a monopoly.

But other forces worked to stir the colonists, particularly the newspaper *Sydney Gazette*, which saw the need of local industries to convert the convict settlement into a prosperous colony of the British Empire. On 7 June 1826, it pointed out that "the people in this colony love to talk and yet do little."

Governor Darling's dispatches in 1823 reveal the same sentiments, but he does mention the firm of Jones and Walker who, it appeared, had five ships and employed over 100 men, and in 1829 reports of a firm, Cooper and Levey, making preparations for a whale fishery at Bennelong Point, close to Port Macquarie, crept into the newspapers.

But it was not until John Bell and Archibald Mosman entered the scene that the industry in Port Jackson really rose on its feet. In 1830, John Bell applied for an allotment of land in Port Jackson, saying: "I am at present fitting up the brig *William Stoveld* for the whale fishery, and not having waterside premises, I am obliged to store and cooper my casks etc. on the King's wharf, and also at this moment I am compelled to pump oil from one cask into another. These operations, viz: the noise of coopering, lumbering the wharf, and the offensive smell of the oil, the Customs House Officers very justly complain of, so much that if I am not allowed to proceed, it will very materially retard my object and affect my interest. In order to prevent a recurrence of the nuisance when the vessel returns, I beg through you to solicit His Excellency the Governor to grant me a portion of unlocated land bounded by the Harbour of Port Jackson, for the purpose of erecting a wharf and suitable premises for the equipment of vessels employed in the whale fishery."

Evidently the Customs Officers were also responsible for the prevention of air pollution at that time, and exerted pressure to have all whaling establishments isolated. When Archibald Mosman, then a prominent merchant and ship-owner, also applied for a similar grant, both he and Bell were given land in Great Sirius Cove. Mosman later bought Bell's share, which gave him 108 acres in what is now known as Mosman's Bay.

Mosman was another who spread his investments, and from his depot in Mosman's Bay sent his ships out all over the Pacific.

In 1846 the following advertisement appeared in Sydney papers:

"MOSMAN'S BAY: The attention of the SHIPPING INTEREST of SYDNEY and the Neighbouring Ports is respectfully requested to the above-named locality. EXTENSIVE Improvements having been effected during the last twelve months, this establishment is now placed in a much more satisfactory condition than at any former period. Vessels drawing sixteen feet of water can float at all times of tide alongside the wharf. The new stone building being completed, there is now ample Storage for 3,000 barrels of OIL; while the accommodation for Ship's Crews, Officers, and Carpenters is in every respect convenient and comfortable.
The HEAVING DOWN GEAR is of the very best description and the premises are placed under experienced superintendence. The natural advantages of the situation

are too well known to require much comment; suffice to say that few artificial wet docks are so thoroughly sheltered from the weather, and the operations of heaving-down and repairing vessels cannot be impeded by the heaviest gales. The distance from Sydney is about three miles.
Masters of vessels who study economy and dispatch are requested to favour Mosman's Bay with an inspection, and to judge for themselves.
Applications to be addressed to Mr Stirling, or to the Superintendent, on the premises.
N.B. Vessels can water at the wharf.
Sydney, New South Wales,
April 21st, 1846.''

Such facilities were a godsend to ships making the long voyage from Britain, but they were also valuable for Sydney whalers, which could be away for well over twelve months. When they returned they could not only be careened but their crews could catch up with their boiling down. And this was the smell to which the Customs officials objected. Boiling down at sea was a foul business, but blubber stored in holds for months at a time must have been overpowering.

By the time Mosman sold his interests to Messrs Hughes and Hosking in 1838, Mosman's Bay had grown into quite a community, but apparently it was not as lawless as the larger settlement, which lay some three miles away. In a letter of recommendation, with which the owners of Mosman's Bay concluded the above advertisement, Captain F. P. Blackwood, R.N. of the ship H.M.S. *Fly* said: "I have much pleasure in saying that I consider your wharf at Mosman's Bay a very eligible spot to careen a ship and heave her down—that we found all the heaving down apparatus in good order, and that the locality would be very favourable to a ship of war from the absence of the temptations of grog &c.; which she would be exposed to at Sydney wharves."

It is hard to imagine Mosman's Bay, full of whalers, being quite as temperate as the Captain suggests, but the mere fact that the facilities existed demonstrated the remarkable progress and stimulus to local industry that whaling had brought to New South Wales. The builders of Mosman's Bay laboured for two years to carve out the area and build the stone wharf out of the rocky virgin scrub which ran right down to the water's edge. No doubt similar facilities would have been built in time, but Mosman foresaw the potential of the large numbers of whalers visiting Port Jackson to boost his trade.

By 1831 cordage and rigging was also being manufactured in the colony and in the same year the whaleship *Australian* was launched; well named, as all her timbers were locally grown and fashioned by colonial tradesmen. On 22 May 1829, the *Australian Newspaper* reported: "Scarcely two years ago, this advantageous branch of commerce (whaling) was confined, we may say, almost solely to one mercantile house in Sydney. Now there are no fewer than thirteen large whaling vessels out of this port." Five years later, this number had increased to forty-two, representing some 10,000 tons of shipping, and in terms of wooden ships this was a figure of which the young colony could be proud. Shipbuilding was also keeping pace with whaling, some thirty-three, representing over 2,050 tons, were launched in 1841. In 1826 whale-oil export from New South Wales totalled £34,850. By 1840 it had jumped to over £335,000.

Not all of this, however, could be classed as produce of the colony or even of its territorial waters, because the Port Jackson whalers, like the Hobart Town vessels, made extensive use of the New Zealand whaling grounds. And it is here, away from all authority save that of their captains, who were as anxious for grease as their crews, that Australian whalers proved themselves as enterprising as their American rivals.

The whales usually arrived at the beginning of May, skirted the western coastline of the North Island, then passed into Cook Strait and Cloudy Bay. By June they were usually at Chatham Islands and down the eastern coast. Others, instead of coming through the strait, headed down the west coast to Stewart Island. Sperm whalers usually favoured the Bay of Islands, on the northern tip of the North Island, while others set up bay whaling stations or whaled off shore for black whales in the bays and inlets of the South Island, and from 1829 onwards small settlements sprang up at Otago, Akaroa, Cloudy Bay, and other favourable spots.

To bring back their cargoes, however, the whalers needed the co-operation of the local Maori tribes. This they bought with supplies of muskets, gunpowder, pipes, tobacco, and rum, and while the natives were systematically butchering one another, the whalers made off with their women (as part of bargains with chiefs) and went about the business of whaling.

In the early years the firms of R. Campbell and Company, Bell and Farmer, and the Weller Bros were very much to the fore, but others followed and the New Zealand grounds became one of the most important sources of oil for Australian whalers.

And so Australian whaling reached its climax; bay whalers in the Derwent, along the Victorian coast, at Twofold Bay in New South Wales, and Encounter Bay in South Australia, and to a lesser degree in Western Australia, added their oil to the spoils of the deep sea whalers from Hobart Town and Port Jackson. The methods adopted were traditional, but it framed its own laws to suit local conditions and gave a much needed stimulus to the Australian economy, by bringing foreign shipping to our shores, and perhaps most important of all, gave Australian shipbuilding a definite purpose.

But Australia was a vast country, and while the whalers worked the sea, others turned to the land; explorers crossed the Blue Mountains and penetrated inland north from Adelaide and across the continent from east to west. Lumbering bullock waggons laying the foundations of our first roads, were bringing in wool from outlying areas, and while some dreamt of using the Murray Darling River system for water transport, others dreamt of finding gold.

Until it was found, there was no quarrel between the land and the sea. But when the first cry of "Gold!" was heard, then the ships, including whalers, were among the first to suffer. As early as 24 June 1851, the *Sydney Herald* made it the substance of a leading article: "The gold fever has already infected many of the crews of the ships in our harbour, and some instances of desertion have occurred from this cause."

The same thing occurred in Hobart Town. Seamen deserted in their hundreds; why should they risk their lives for a miserable lay when they might find a fortune on the goldfields, be their own masters, and perhaps set themselves up for life? Those who did stay in the seaports went from ship to ship, selling their services to the highest bidder.

But one of the most serious results of the rush was the shortage of tradesmen in the whaling ports; sailmakers, rope makers, coopers, and shipwrights; these men, who had worked behind the scenes for so long to keep the whalers at sea, were no longer available. Even the "lucky" captain of a greasy ship faced months of idleness when he returned to port, while his agents scoured the colonies for men and equipment.

By 1853, exports of whale-oil from New South Wales had fallen to £16,000. In that year over £1¾ million worth of gold passed through the dealers. The *Herald*, in its wisdom, was moved to issue a warning, almost a prayer for sanity during this time of crisis: "Exaggeration would be as cruel as it is unnecessary. But it becomes the duty of every sober-minded man in the community to look the danger calmly but fully in the face. That there is gold on the surface of our western interior is a fact which cannot now be doubted. But let us cling to the hope, unless driven from it by irresistible evidence, that the treasure does not exist in large quantities; that the cost of finding, collecting, and conveying it to the market will prevent the speculation from being more than moderately remunerative; and that experience will soon convince the masses of the people that, after all, the ordinary pursuits of industry are the safest and best. Should this hope be realised all will be well . . . but should these hopes be disappointed—should our gold be abundant in quantity, rich in quality, and easy of access—let the inhabitants of New South Wales and the neighbouring colonies stand prepared for calamities far more terrible than earthquakes or pestilence."

Their worst fears were realized, because the gold was abundant in quantity, and the whalers would have been the last to agree that "the ordinary pursuits of industry are the safest and best." In other articles of the day the *Herald* described many who joined the rush as "men who would hesitate to walk the length of George Street in a shower of rain," and "people of all trades, callings and pursuits were quickly transformed into miners, and many a hand which had been trained to kid gloves or accustomed to wield nothing heavier than the grey goose-quill became nervous to clutch the pick and crowbar," or "rock the cradle at our infant mines."

It would be interesting to know the number of whalers who made a success of gold mining. Their hands had certainly clutched heavier instruments than grey goose-quills, and they were certainly as well or better equipped physically than the average shop keeper, clerk, tailor, or candlemaker.

The element of chance probably appealed to many of them; fortunes could be made if "mother luck" smiled on miners in the same way as a whaler could fall in with a big school of sperm whales at sea, or be blessed with the arrival of large numbers of right whales at some lonely bay whaling station.

Unfortunately, the whalers lost their identity as soon as they joined the thousands on their way to the goldfields, and the whole crowd was simply described by contemporary writers as miners. No records exist which tell of the individual success of any whaler turned digger.

But even before the discovery of gold there were other disturbing signs within the industry. These were to have an even greater effect on its future, particularly on bay whaling. John Dixon, while trying to enlighten intending immigrants on their future in Van Diemen's Land, wrote in 1839 that: "The commerce of Van Diemen's Land is insignificant; and its trade with the mother country narrow and trifling. It exports wool, oil, and whalebone; and sometimes a little bark. Wool is its primary production, the staple commodity of the colony. Its fishing is precarious, yet it is prosecuted with vigour and enterprise. Seals were numerous here some years ago, but these have now all disappeared. And the whale, too, that one time was so attached to the coast, seems now to be deserting it; for, year after year, the whaler has to seek it at a greater distance."

Despite such observations the bay whalers seemed to see profitable times ahead, and as late as 1851 they were still making applications for land to set up bay whaling stations around the Tasmanian coast.

Meanwhile, other naturalists were issuing warnings both in London and Hobart Town based on the numbers of right whales taken in the New Zealand grounds. They had done the same thing regarding sealing, but their opinions and warnings fell on deaf ears until, as John Dixon had pointed out, the seals disappeared altogether.

It is hard to imagine how a few thousand whalers, using primitive methods, could appreciably affect a population of right whales which for centuries had migrated to the northern breeding grounds. And from scattered records it is difficult to determine the numbers killed each year. E. J. Slijper, Professor of General Zoology at the University of Amsterdam, puts it at approximately 14,000 a year in the first 20 years of the nineteenth century. A great proportion of these would have been cows and calves as they were always the easiest to hunt, and this was probably the reason why by 1843 bay whaling had begun to decline and within a few years ceased altogether. Once the whalers had begun their onslaught upon the "breeding stock" of the whale population, then there would be a rapid and progressive decline in its numbers.

This decline served to spur the deep sea sperm whalers to even greater efforts, and between 1820 and 1850, the heyday of sperm whaling, an estimated 70,000 men from America, England, France, and Australia killed approximately 10,000 sperm whales a year. But here the picture was a little different. Sperm whalers roamed all over the Pacific and the killing was spread over a much wider area and not confined to definite breeding grounds. This, no doubt, saved the species from extinction.

But even during these years disturbing features began to develop. In the early 1800s, whalers putting out from American and British ports could count on filling up within two years. By 1844 it took anything up to 4 years. Australian whalers were faced with the same situation, but with cheap labour and water and food available in the Pacific islands, this did not cause great concern. In the end, it was not the decline of the whale population but the exploitation of another of this planet's natural reserves which brought sperm whaling from sailing ships to an abrupt end. In 1859, mineral oil was discovered in Pennsylvania. By 1861, over two million barrels were flooding the market. This was only one step from kerosene, which ousted whale-oil as lamp fuel and brought the sperm whalers to their home ports for the last time.

Between 1866 and 1890 the value of whale oil on the American market dropped from \$2.75 a gallon for sperm-oil and \$1.25 for whale-oil to 66 cents and $4\frac{1}{2}$ cents respectively.

There was, of course, still a small local market; sperm candles sold at 2/- a pound, and sperm oil at 4/- a gallon, in Port Jackson in the good days of whaling. But as early as 1841 the city became the first in the southern hemisphere to be illuminated by gas light, little more than twenty years after it had been introduced into the leading provincial towns of England.

Australia was no longer an isolated colony at the other end of the world, and many whaleships plying between the New World and the Old had first brought the news of achievements which were to cripple the very trade which kept these ships at sea.

And how eagerly these new achievements were seized upon and put to use in Australia. The *Herald* reported gleefully that:

It was a memorable day—or rather night—when the offices of the *Herald* were first lit up with gas. This may, to those who are not printers, seem a small matter. To the printer—especially the compositor—it meant a great deal; something more than a better light for his work. The old printer's candlestick—a tin tube let into a base about an inch and a quarter square and fitted with a bar to keep it from toppling over—with the 'slip' candle, disappeared forever. And with them went the snuffers—always in the way—and the streams of grease that frequently ran down

Blue whale cow and calf, taken only 300 yards off Leighton Beach, in Western Australia (*W.A. Newspapers*)

"Fast fish!" A whale-catcher off Cheynes Beach, Western Australia, heaves-to after a whale has been killed by her harpoon-gun (*Cheynes Beach Whaling Co.*)

into the boxes of the 'lower-case,' causing delay and vexation to the compositor and loss of time to his employer. It requires a man to experience the change to completely appreciate it. A little calculation will show the loss of time. Each man was provided with two candles, with which he worked, say, six hours. The snuffing operation had to be performed twelve times on each pair of candles, a task which would occupy at least fifteen seconds. That is three minutes. There were about thirty pairs of candles lit each night; and the loss of time, therefore, totals about one hour and a half a night, or nine hours per week. It isn't much, certainly—but run through the year, at the wages then paid of 8/- per day, and it tots up to something over £13 a year in lost time.

That sum could have bought the services of a whaler for much longer than a year, but the article reflects the mood of the times.

A great deal had been accomplished since Captain Melville had waited on Governor Phillip and been given the green light for his whaling ventures. And although the Australian colonies had passed through periods of what were described as depressing times economically, their futures were assured.

The decline of the whaling industry, although it meant dwindling exports and caused a definite setback in the shipbuilding industry, particularly in Hobart Town, and the closing down of the careening docks at Mosman's Bay, was to be a passing phase. The discovery of gold and the growing success of the pastoral industry pushed it into the background.

There were, of course, a few diehard captains who had spent a lifetime at sea and who refused to admit that deep sea whaling from sailing ships had no part to play in an era which now boasted cotton material, gas, smokeless kerosene lamps, and steam engines.

One was Captain W. Folder who, as late as 1890, set sail in the converted clipper-barque *Helen* from Hobart for Campbell Island, south of New Zealand. Here he rendez-

The Dutch whale-factory ship *Willem Barendsz* (*P. P. Hey*)

A Dutch whale-factory ship lies hove-to amongst drift-ice of the rich Antarctic whaling grounds (*P. P. Hey*)

voused with the *Southern Cross*, of the Newnes-Borcherevink Antarctic expedition. The arrangement was that the *Southern Cross* should take whales, and tow them to the *Helen* for trying out. They met with moderate success, but the most significant factor was that the *Southern Cross* was a steamship. A new era had begun, in which the whaler would be no longer at the mercy of wind and tide, and would not have to grapple with the whale almost at arm's length. It was to be the era of mechanized slaughter, which would devastate the herds of whales which were slowly building up their numbers after the lull in the hunt.

The old ways were passing, and new inventions, which would eliminate the wastage of primitive methods, were on their way. These would eventually call Australian whalers back into the industry.

But before joining the new breed, it is appropriate to say a final word or two about the pioneers.

During the current educational ferment, various writers are taking a fresh approach to the writing of Australian history, and if one assumes that the purpose of teaching history is partly to acquaint younger students with the activities, motives, circumstances, and ideals of our pioneering forefathers in the march towards nationhood, then the whaling industry and the men who pursued it deserve a more prominent place.

Most primary students follow the dismal trek of Edward John Eyre across the Nullarbor Plain, and are aware of the reasons for his journey, but the fact that he was rescued by whalers who were growing food on a remote part of the Australian mainland before he left Adelaide is dismissed in a few lines.

Students learn in detail the trials and tribulations of Captain Sturt's voyage down the River Murray and many believe he was the first man to examine in detail the Murray mouth. And yet the same area was familiar to whalers and sealers from Kangaroo Island, at the foot of St Vincent Gulf in South Australia, years before he launched his whaleboat in the upper reaches of the Murray/Darling river system.

The coast of Victoria, the islands of Bass Strait, and the

shores of Tasmania were explored by whalers and sealers, and whalers arrived at Encounter Bay in South Australia before the first organized settlers.

The average student, if asked to name the creature on which Australia's first industry was based, would almost certainly say that it was the sheep. He knows more of the exploits of John MacArthur than of Captain Thomas Melville, and yet Melville pioneered an industry which, as far as exports during the first thirty years of N.S.W. were concerned, was just as important as wool.

Hobart Town was established as a convict settlement, and the student is quite familiar with prison life at Port Arthur, but it was whaling which turned it into a world famous deep-sea port and supported its thriving shipbuilding industry.

Whaling played an important part in the formation of the South Australian Company, which brought the first settlers to South Australia, and it was a whaler who circumnavigated Tasmania in an open boat. And yet students are more familiar with the deeds of Ned Kelly than they are with those of Captain James Kelly, who faced as much danger in his forty-nine day voyage around the Apple Isle as Captain Sturt did on his voyage down the Murray.

It can be argued, of course, that whalers and sealers were unofficial explorers, without government backing, who found themselves in remote places and in many cases did not keep "official records." Therefore they may perhaps be regarded as unreliable, and only deserving of a small place in our recorded history.

On the individual level there is some truth in this. As with the exploits of our wandering stockmen, it would be impossible to include in a new treatment of Australian school history every daring act and every life sacrificed by these men in the whaling industry.

And yet collectively the exploits of our whalers were a part of "the great reaching out" which filled in the outline of our coast and put dots on our remote interior. Land explorers set out in search of new pastures and settlement followed. Whalers reached out in search of new whaling grounds and settlement followed. Captain Kelly discovered and explored Port Davey and Macquarie Harbour in Tasmania.

Whalers, unaided by cheap convict labour, established an industry in competition with overseas interests, and ships built in Australia, to serve that industry, formed the nucleus of our coastal and overseas fleets.

The merchants, shipowners, and investors who sank money into whaling did as much to put Australia on its feet as those who put money into pastoral properties.

The importance of all this was often overlooked by our early writers. In the fashion of the day, and in the best Anglo-Saxon tradition, the element of competition involved made them liken the exploits of our early whalers to a grand form of sport: "Currency lads partial to the sport," and, "at the same moment, we are rearing up a fine and manly race of native youths, in a nursery that would qualify them to contest the palm of superiority on the water, with the inhabitants of any country upon the whole face of the globe."

This "grand form of sport" style of reporting often crept into despatches concerning naval and land battles, and readers were spared the unpleasant truth about casualties among the lower ranks until the strategy which brought victory had been driven home.

But in fact there was no more voluntary sport in early Australian whaling than there was in soldiering, and like the men who did the actual fighting to ensure that the sun never set on the British Empire, Australian whalers played an important part in our history.

In the entertainment field, Australian film makers, novelists, and T.V. producers have also given the old whalers scant attention. Only one book of merit, *Paving the Way*, centred around the whalers at Encounter Bay in South Australia, has been written, and yet there is a wealth of excitement and human tragedy in the log books of Australian whaling ships.

One explanation for this is that anything written or produced could only be a poor imitation of *Moby Dick*, but the Australian whaler was no imitator; his methods may have been traditional, but he adapted himself to his environment, framed his own laws, and worked along a sea coast which was amongst the loneliest and most dangerous in the world. Like the exploits of our travelling stockmen and drovers, which equal the adventures of the American cowboy (if one discounts the mythical stories which have debased the genuine achievements of that breed of men), the era of Australian whaling produced characters more romantic and genuine than many of those given a phoney immortality by overseas writers.

10

Twelve tons of oil apiece

"You often hear people refer regretfully to the 'dear old days of romance'—to the time when the knight-errant roamed the earth to do honour to his lady's eyes. These folks complain that we are living in an age of realism! An age of realism? Why, this is the most romantic of all ages! An age when the human voice is hurled across the world without wires; where the temperature of Mars is taken from more than thirty millions of miles away; when tons of steel carrying precious human lives ride easily and safely through the air or under the sea. The advertising columns of this paper are full of this modern romance—stories of things produced by men who have devoted their lives to bringing new comforts, conveniences, and pleasures to mankind. Advertisements tell of their achievements not with the exaggeration of a jongleur, but with calm sincerity. Here is a firm that has spent a fortune to develop a product that makes your baby more comfortable. Here is a company that has laboured fifty years to cut a single hour of toil from your day's work. Here is a man who has searched the seven seas to produce a new flavour for your dinner. Romance? This age is full of it. Read the Advertisements. They tell you what the magicians of industry are doing for you." (Adelaide *Advertiser*, 17 June 1933).

The magicians of industry in the early 1900s, while working purely for the benefit of mankind, had not overlooked the whale, which had lost most of its economic importance when earlier magicians developed gas light, electricity, and mineral oils.

While their researches into the potential of the whale's carcass may not have taken an hour of toil from a day's work in the factory, they could certainly take more than an hour off a man's life. Glycerine, used in the manufacture of explosives for the first world war, was a by-product of the whaling industry. But in all fairness it must be said that they did use the whale to make baby more comfortable, and meals easier to prepare.

"South for whales. Hopes for kill worth £4,000,000. What is expected to prove the greatest slaughter of whales that the world has ever known has just begun. It is estimated that 25,000 whales will have been killed when the season ends in March. A fleet of 129 vessels, of which 17 are mother ships and 112 whale chasers or harpoon gun steamers has gone from Norway to South Georgia. Seaplanes will be used to spot the whales and the vessels carry wireless. The whales are struck by harpoons attached to steel ropes and shot out of a gun. Once harpooned the whales are killed by explosive bombs. They may even be electrocuted by a current sent along a wire . . . What will become of this flood of oil? It will be used very largely in the making of soap, margarine, lard, and other substances which most people fail to associate with whales."

This great revival described in the Adelaide *Advertiser*, 1 October 1932, did not come about overnight, nor did Australians play a leading role as they had done in the early days. In fact they let a golden opportunity slip through their fingers, although new whaling grounds were discovered within a few days of sailing of the Australian coast.

To understand the era of modern whaling and the conditions which made it profitable (for other countries) it is necessary to go back to the "dear old days of romance" when whalers "searched the seven seas" for oil to light the lamps of European cities.

During the twenty blank years of deep sea whaling which began in 1875, when whalers all over the world were forced into a retirement from which many of them would never return, the whale population of the world was left to increase. The sperm, and to a lesser extent the right whale, evidently made the most of this opportunity, and towards the end of last century were once more seen in fairly large numbers in the oceans of the world.

Two species of whales, the blue and his close cousin the finner, had lived almost unmolested during the previous era of whaling. Because of their speed, fighting ability, and great strength, they had been given a wide berth by whalers in open boats. One can appreciate the whalers' caution. Blue whales have been known to tow small steamers through the water after the order "full astern" had been given. They could have towed a thirty-six foot timber whaleboat for as long as the line held, without the slightest effort.

Whaling articles are full of references and quotes about their size and strength and most bear out the old whalers' saying that "a whale weighs a ton a foot." The Adelaide *Advertiser* of 1 October 1932 stated that: "Blue whales average twelve tons of oil apiece, finners seven, and humpbacks five. In recent years blue whales have made up a very large part of the catch in the Antarctic and sub-Antarctic seas. It is roughly reckoned that a blue whale weighs just more than a ton for every foot of its length, and one killed off South Georgia was 107 feet long."

But even while blue whales and finners were enjoying immunity, events were taking place which were to make nothing of their speed, strength, and courage.

The first had taken place in 1802, when William Symington fitted a steam engine to a small boat and successfully tried her out in the Forth and Clyde canals in Scotland. By 1819 the first auxiliary steamer had crossed the Atlantic, and eventually the screw propellor was forcing ships through the water at twenty-one knots.

While the rivalries and jealousies between men of sail and the new breed of steam sailors rose to fever pitch, the whalers continued working. Although steamers were a commonplace by 1875, the whalers did not at first see that they offered any answer to the problems of their particular trade. They knew that to kill a whale a harpoon had to be sunk into its hide. A boat could be lowered from a steamship, but muscles were needed for the kill. Apart from this, fresh whaling grounds had to be found.

The first step in this direction, however, had been taken even before deep sea sperm whaling was at its height. In 1839, an English naval captain, James Clark Ross, called at Hobart Town on his way into Antarctic seas. He drove his ships south through the ice-packs until he discovered a great ice-free bay in the Antarctic continent. Now known as the Ross Sea, it is some 2,000 miles south of New Zealand. Here he reported seeing right whales in their thousands.

His reports were well noted but put aside by whaling captains. They knew that their crews would accept the perils of being thrown into the comparatively warm waters off the coasts of New Zealand and Australia and in the Indian Ocean, but it would take more than a small lay to encourage them to face the icy seas of the Antarctic, where a smashed boat could mean death within a few minutes.

Many a whaling captain cursed with a dry ship must have been tempted to set a southerly course, but before this could happen a means of delivering a harpoon into a whale from the safety of a larger ship had to be found. The obvious answer, of course, was a harpoon fired from a gun with a range of fifty or sixty yards. Ships could be blown apart with cannon balls, but it was not so easy to fire a long, pointed missile trailing a considerable length of strong line.

Many attempts were made before a Norwegian whaler brought his experience to bear on the problem. This was Svend Foyn, often described as "The Grand old Man of Norwegian Whaling." He was born on the island of Notteroy near Tonsberg, Norway, in 1809, and during the heyday of Australian whaling he was employed on the other side of the world, in the Scottish sealing trade in the Arctic. He made many voyages after both seals and whales, and these gave him plenty of opportunity to study the dangers and inefficiency of whaling from open boats. Although he became a successful whaler in the old-fashioned style, he was never satisfied with the hit or miss methods then in use, and devoted a great deal of time and money to perfecting a weapon which would secure the biggest and most agile whale. It was his harpoon gun which was to bring the blue and finner whales within reach of the hunters, and give new impetus to the whaling industry.

His first harpoon gun, invented in 1864, weighed a ton, and was to be mounted on a steel swivel in the bows of the whaleboat, so that it could be elevated or lowered while moving around in a complete circle. It had a four foot barrel, a three inch bore, and was muzzle-loaded with an immense charge of over 300 pounds of black powder, pounded down with a ramrod. A propellant as powerful as this was needed to shoot the harpoon, and its trailing length of line, with sufficient force to penetrate the whale.

The harpoon consisted of a shaft which fitted into the barrel, with a barbed head containing an explosive charge and a small vial of sulphuric acid. When the harpoon hit the whale the vial broke, and the acid escaped setting off the charge. The explosion caused the arms to fly open thereby fixing the harpoon more firmly in the flesh.

However, it was found that his invention was several years ahead of its time. It was out of the question to install such a cumbersome piece of equipment in the bows of an open whaleboat, because the firing blast and recoil would have sent all hands to the bottom, while steamships that were big enough to carry it were too slow for whale hunting, and could not manoeuvre with sufficient speed. What was needed was a small fast steamer, which could approach within fifty yards of a blue whale.

Such vessels appeared in 1880, in the form of whale chasers, thirty feet long with a beam of thirteen feet and a nine foot draught, and an engine which pulsated with the necessary power. Their launching was to send the remaining open boat whalers to port for the last time, and usher in the modern period of world whaling which eventually spread to the Australian coastline. Before this could happen, however, they had to be tested in action.

They were put into service along the Norwegian coast, and were so successful that the new system of whaling, incorporating Svend Foyn's harpoon gun, spread within a few years to Iceland, the Faroes, the western coast of Ireland, and Newfoundland. In 1897, over 2,000 whales were taken off Newfoundland alone. The Norwegians had re-established shore based whaling with a vengeance, and capital flowed into the industry as fast as it had done in the sailing ship era.

But before long, the Norwegians were faced with the old problem of a dwindling whale population, and were forced to seek new grounds. Their attention was immediately focused on the southern seas and in 1891 Carl Anton Larsen set sail for the Antarctic. A Scottish fleet, under Captain Robertson, sailed at the same time, but on arrival they both found that the right whales seen by Ross had disappeared. Instead, they saw blue whales.

But the fleets were not equipped for these monsters and had to be content with seal skins and seal blubber. Captain Larsen made several more voyages, and in 1922 called at Tasmania with a small fleet consisting of a 12,000 ton freighter and five small whale chasers.

He converted the freighter into a floating factory ship, which succeeded in towing the chasers in line through the pack ice and into the Ross Sea.

The harpoon gun had done away with hunting from open boats, but the whalers had to face a new hazard—ice. Cutting up the whales alongside the ship was often impossible during gales, but despite this they took 200 blue whales. The modern era had begun.

One problem, however, still remained; getting the carcasses on board for trying-out. This was overcome in 1925, when factory ships were built with a huge slipway at the stern through which whales could be hauled aboard by steam winches.

The Norwegians must go on record as the pioneers of Antarctic whaling and by 1930 they had twenty-three companies using thirty factory ships and 145 whale chasers in Antarctic waters.

New inventions brought new terms and new rates of pay for whalers. The Adelaide *Advertiser* of 1 October 1932 reported that: "A harpoon-gunner named Jorgensen, working on a chaser belonging to the *Sir James Clark Ross*, killed 245 whales in the 1930-31 season in the Ross Sea. His earnings for the voyage—harpooners are paid so much for each whale killed—were £3,000."

The Germans, the English, and the Japanese followed the Norwegians into southern waters and in the 1934-5 season some 40,000 whales were killed and processed.

The methods of hunting and handling the whales were almost completely mechanized by that time, and compared with the old style of trying out on a lonely beach or on the deck of a sailing ship were like a 1,000-watt electric lamp against a spermaceti candle.

First, the whale was hunted down by one of the fast, manoeuvrable chasers, and killed expeditiously by the explosive harpoon. It was then marked with a flag. But before the chaser steamed off in search of further prey, a hollow lance connected by a hose to a compressed air pump was jabbed into the carcass, and the whale was blown up like a giant balloon. This innovation overcame another of the problems faced by old whalers. On the rare occasions when they did manage to kill a blue or finner, they often had to stand by, helpless and frustrated, and watch it sink to the bottom.

A wireless message to the factory ship, giving the dead whale's position, was then sent out before the chaser left the carcass to be towed back by the buoy boats.

When the tow was completed, the carcass was moored tail first at the stern of the mother ship, where it was grabbed by an enormous claw and hauled up the slipway to the enormous deck, as big as a football field. There, in comparative comfort and away from sharks and killer whales, the carcass could be butchered.

Primary cuts were made by the head flensers near the head, and backwards for the whole length of the body and along

the sides. A hook was then secured, the winches operated,
and the whale peeled like a banana.

Head flensers were also responsible for cutting away the
roof of the mouth, once the storehouse of valuable baleen
or whalebone. This was now thrown overboard. While this
was being done, junior flensers hacked the blubber into strips
about two feet wide and ten feet long, which were then
attached to metal hooks by the blubber-boys, and towed to
open manholes on the deck.

Below decks, the blubber passed through revolving
choppers before sliding into the cookers. These are another
invention of the "magicians of industry"; a gigantic pressure
cooker working on the same principle as the domestic model,
which can be regulated for various grades of oil. Centrifugal
force is then used to separate the waste from the pure oil.

While this was proceeding, the remains of the carcass were
being treated by "lemmers," who cut away at bones and
meat with power saws. Practically the whole whale finishes
in the cooker, only a few parts of gut being thrown overboard;
a far cry from the days when men were satisfied to ladle a few
barrels of oil from an open trypot.

Today the main products of the industry are lubricants,
margarine, soap, polishes, fertilizers, vitamin A from the
liver oil, insulin from the pancreas, and hormones from
the pituitary gland.

But the Japanese who followed the Norwegians and
Russians into Antarctic water, saw added attractions in
whaling and with their usual thoroughness made the most of
every opportunity. The Japanese Whaling Association,
Fishing Agency of the Ministry of Agriculture and Forestry,
sums it up by saying:

> Japanese whaling expeditions to the Antarctic began in
> the 1934-35 season with one factory ship and three
> catcher boats. It developed rapidly and culminated in
> 1938-39 when six fleets and forty-nine catchers participated
> in Antarctic whaling. Since that season these six fleets
> continued their Antarctic whaling until the World War
> forced the suspension of the operation in 1942. All these

factory ships and a number of catcher boats were lost in the course of the war. Immediately after the termination of the war, the Japanese whaling industry had to make a desperate effort to resume Antarctic expeditions in order to save the country from the serious food crisis. Differing from fleets of other countries which principally operate for whale-oil, the Japanese fleets place great importance upon the meat. The Japanese factory ships, therefore, are accompanied with a number of refrigerating ships which freeze the meat and transportation ships which carry home the frozen or salted meat. Besides the meat, various parts of the whale are eaten or utilised other than for food. The blubber residues, after the oil is removed, are the favourite materials for stew or soup. Salted tail flukes are sliced or cut into noodle form and treated with boiling waters to get conglutinated protein which is eaten being mixed with soy-paste. Cartilage from the upper jaw bone is eaten after being pickled.

Among non-food products of whales, cigarette holders and art goods are made from the teeth of sperm whales. Various fancy wares are also made from baleen plate. The bones, after oils are extracted, are crushed and used as phosphatic fertilizer. The crushed bone is also used as potter's clay.

And so even a country defeated in war joined those from the northern hemisphere, and made whaling profitable in the waters south of Australia. During the period under review there is no mention of a revival of the Australian industry, simply because it was non-existent. On 1 October 1932, the Adelaide *Advertiser* said: "The Norwegians have almost a world monopoly of whaling. It is more than thirty-five years since any Australian owned vessel went awhaling."

The Norwegians, who pioneered the harpoon gun and the factory ship, had to land on our shores and re-introduce shore stations before Australians again harpooned a whale.

11

Old occupation to be revived

"A BARREL ATTACHED TO A LENGTH OF SMALL LINE WAS placed over the side. The vessel was stopped and when the barrel was about 100 feet distant Captain Olsen fired. The harpoon hit the target just on the underneath portion. It was a well-delivered shot, and could not have failed to prove fatal to the whale had there been one in that particular spot."

This demonstration was made by the crew of the Norwegian whale chaser *Klem* at Fremantle, towards the end of June 1912. She was one of two chasers attached to the 3,250 ton factory ship *Vasco da Gama*, of the Western Australian Whaling Company.

The Norwegians had been whaling on the western coast of Africa in 1911, and in their search for new grounds had gained licences from the Western Australian Government for the northwest coast.

Eager to impress, they not only demonstrated their harpoon guns, but held a reception on board the *Vasco da Gama* for members of parliament and leading business men.

Their aim, however, was not to raise capital in Australia. This was readily available in Norway, where reports flowing

115

in from countries beneath the equator caused something of a whaling boom. Some Norwegian groups sent captains out to assess the situation, while others gathered information from retired Australian whalers, State Government Departments, and New Zealand authorities.

Some of this information may have been correct, but if it was, they must have been told that the right whale was almost non-existent, and at no time had there ever been large numbers of blue whales in our waters; and that even the sperm had suffered serious depletion over the years.

Despite this the Norwegians were enthusiastic. They saw a continent stretched right across the southern ocean, a natural barrier and haven in the northward migration from the Antarctic, a base of operations from which, if necessary, they could sail south to meet the whales which Captain Larsen and others had seen in the distant Antarctic waters.

They were also no doubt stirred by reports of the existence of humpback whales, particularly on the northwest coast of West Australia.

Why this particular species could still be found in large numbers in 1912, when its cousins, the right whales, had been hunted out, is an interesting point.

There are, however, five self-contained humpback populations in the southern hemisphere, two of which migrate to Australian waters; one group to the east coast of New Zealand and Australia and the islands of the south-west Pacific, and the other to the Western Australian coast.

The western coast family had some refuge in the more northerly waters, where the inhospitable shoreline was not as inviting as the eastern coast of Australia for the shore based operations in the nineteenth century.

Apart from this, the humpback was not quite as lucrative as far as oil was concerned, and if given a choice the old whalers made it a second target. And so they survived to bring whaling again to Australian waters.

The Norwegians arrived at an opportune time, when the northwest coast was much in the headlines. In July of 1912, the then Minister of Works, Mr W. D. Johnson, submitted to

An automatic grab is lowered to heave a whale aboard the whale-factory ship *Ulysses*, during the 1937 whaling season (*Library Board of Western Australia*)

Cutting-in a whale aboard a factory ship—a vastly different process from the dangers and discomforts of cutting-in to a whale secured alongside one of the old-fashioned whalers (*P. P. Hey*)

cabinet a report on his investigations of the area, saying in
part:

> The whole coast is in a bad state of disorganisation.
> The methods are so crude and out of date that one cannot
> understand how the various ministers could go along and
> inspect them and allow such a sad state of affairs to
> continue. One would be justified in saying that 33 per
> cent of the expenditure on the coast has been wasted
> through want of proper supervision and lax administration.
> This area is full of possibilities and it is questionable
> whether the whole administration of the coast is not
> worthy of being placed under the control of one minister
> who would be responsible for its general development.

The government, eager to open up the northwest area and
no doubt impressed with the efficiency of the Norwegians
(the harpoon gun and the factory ship were certainly not
crude and out of date), gave them every encouragement at
the time. The Western Australian Whaling Company, which
was floated in Norway with a capital of £100,000, was
granted the exclusive rights for seven years to operate
between Steep Point, near Dirk Hartog Island, and Cape
Lambert. Whales of the humpback variety were said to be
plentiful along the nor'west coast, and it was estimated that
a steamer should, during a fair average season extending
over six months, capture 150 to 200 whales. The value of each
whale, at that time, was something like £100.

While the seven year term was quite generous and the
Western Australian Government was eager to see any new
moves made in the great empty spaces of the north, they were
also anxious to see new place names on the long line of their
coastal maps, and insisted on an agreement which would
bring the visitors ashore on a permanent basis. The Annual
Report of the Chief Inspector of Fisheries of Western
Australia, published on 4 January 1914, said:

> Under the terms of exclusive whaling licences, as
> granted in this state, it becomes necessary for those
> concerned to within a certain period, erect a station or

Herman Melville called them "meat pies a hundred feet long." A blue whale
waiting for cutting-in. (*P. P. Hey*)

Sperm whales lie on the deck of a whale-factory ship. The funnel of a whale-catcher
can be seen alongside (*P. P. Hey*)

stations for the treatment of the carcases—after the blubber
has been removed—of the whales taken, and as far as
practicable the products of such treatment such as
fertiliser, bone meal, etc. are to be disposed of within the
state. When we take into consideration the fact that each
carcase will roughly produce four tons of excellent
fertiliser, worth about £7 per ton, and during one
successful season's fishing about 300 whales would be
taken, it is obvious that once these stations become
established, not only the company but the state is likely
to reap a considerable benefit from this industry.

Although it was their original intention to erect shore
stations, it is unlikely that the Norwegians would have
concerned themselves with the manufacture of a great
quantity of fertilizer in the early years had the Western
Australian Government not brought pressure to bear. The
equipment was costly and had to be brought from Norway,
and it would have been a much better business proposition
to put the companies on a profitable basis before over-
extending themselves.

They could have operated from temporary sheltered shore
bases; after all, this was the purpose of the factory ship.
The chasers went out after the whales, made the kill, towed
them back to the factory ship following in the background,
and if the weather was calm they could be stripped of blubber
and the carcass then either set adrift or if necessary towed
back to shore for fertilizer treatment later.

The fleets only required temporary shelter in the event of
rough weather and for supplies of food, bunkering coal and
water.

At this time the slipway in the stern of the factory ship
had not been introduced, and the Norwegians were using
small punts which were lowered over the side for the initial
cutting up.

The *Vasco da Gama* was the first factory ship to visit
Western Australia, and excited considerable interest. An
eyewitness report said that:

The vessel, which is about 3,250 tons register, has been

fitted out with all the latest appliances for the treatment of whales and reducing them to their saleable products. Her lower holds are fitted with immense tanks fore and aft, these being the receptacles for the oil after it has been extracted from the blubber and the flesh. Most of the deck spaces are taken up by steam boilers while quite a large portion of the for'ard deck is devoted to a blacksmith's shop. . . . The residue from these boilers is subsequently utilised as fertilizer and cattle food. On the *Vasco da Gama*, however, there is not sufficient room for the drying treatment and milling of the residue. Once land stations have been established along the nor'-west coast the residue will be converted into fertilizer and so forth.

But Western Australia, and in particular the great wheat belt, was in urgent need of fertilizer and the arrangement was a satisfactory one for the State, if not quite as convenient for the Norwegians. This was made clear in a somewhat disgruntled comment in the 1914 annual report of the Western Australian Whaling Company:

It cannot be denied that the doubtfulness of the company's future that had made itself felt at the starting of our operations, interfered to a great extent with our work; most of the circumstances that brought this about should not have been debited to the company. Even when it is proved that the whales are here and the difficulties with the harbour and water are overcome, we must be prepared for the fact that some time will have to expire before the company is fully organized and can be said to be in proper working order. As the company has an agreement with the Government of Western Australia to build guano factories, for reasons stated before, we would not send our land station out. Therefore our representatives in Western Australia managed to get an extension of time in which to build these factories, but it cannot be for long, and this large expense will then have to be made.

The representative mentioned was Mr August Stang, the

Norwegian Consul in Perth, who worked tirelessly during this period to establish the industry. He later became General Manager and Attorney in Western Australia for the Christian Nielsen Company of Larvik.

This company was described by Mr Stang as "a company of General Managers of whaling companies responsible for the three firms operating in Australia." When questioned before a select committee to enquire into whaling in Western Australia in 1915 he described the set up as: "Separate companies with separate shareholders. They certainly have the same manager who is paid so much per year office allowance, and who is further paid a percentage of the profits when the profits materialize, but they are certainly not a combination."

This grouping of three companies under one manager, all operating at times in the one area, managed to confuse everyone except the Norwegians who, through long experience, knew the advantages of group whaling for the mutual protection of the grounds from outsiders.

Apart from the cost of establishing their guano factories, the Norwegians were faced with several other difficulties.

Firstly they had to establish the existence of whales in commercial quantities and then find a sheltered harbour for their factory ship within close range. Fresh water was also a problem on an uninviting lonely shoreline. This was overcome by sinking coastal wells, an unpleasant task for men from a temperate climate. Fresh water had never been a problem in Newfoundland but in Australia it was number one priority. The new era of whaling based on the factory ship and the harpoon gun still had its share of land work.

Food supplies, bunkering coal, and oil barrels were also needed, but these could only be obtained in Fremantle and required a shuttle service between their nor'-west station and the south.

In a further statement before the enquiry committee, Mr Stang said that the companies had spent £80,000 on coals, provisions and general supplies, and £10,000 on exploratory work. He also stated: "Commonwealth and state departments have a singular facility for getting money.

I have paid between £6,000 and £7,000 apart from licence fees for customs and railway freight." The licence fees for the three companies came to approximately £1,000 and on top of this £30,000 had been spent on the shore station at Frenchman's Bay.

And so the revival of Australian whaling brought foreign ships to our ports and gave a lift to our economy, but this time on a much smaller scale. No cordage, sails, shipwrights, and barmaids were needed to keep the ships at sea, but for all that, the impact of the Norwegians was exciting, deserving more co-operation than they received from the authorities and public at large.

For most Australians, whaling was dead and buried, only revived by an ageing few who could remember it as exciting stories for their grandchildren. The old whalers had taken their place alongside our pioneers, and the factory ship and harpoon gun, while creating a ripple of interest, was seen as something entirely new, completely divorced from the past.

While they were struggling, the companies seldom made news paragraph. It was only when they appeared to be succeeding, and had to take steps to protect their interests, that they were raised to the dignity of an enquiry by a select committee, and caused rumblings in the trade union movement.

During this period there were those who wished to see Australian waters kept for Australians, but although some licence applications were received, no very definite moves were made to launch an Australian whaling company.

On paper, at least, it would have required far less capital than the Norwegians put into the industry. There would have been no need for a factory ship, because the chasers could have operated successfully from a shore station, as they did later. The boiling-down procedure was known, and there was a ready-made home market for fertilizer and an export market for barrelled oil.

But the whole business then centred on the quick and sure killing of a number of whales made possible by the harpoon gun. The Norwegians pioneered the successful use of this piece of equipment, and held the patents. Now that they had been granted licences to fish our waters, it is unlikely that they would have supplied any technical assistance to competitors. They were the first in the field and were in a position to take the initiative. They were not slow to move, and a report in the *Western Mail* of 6 December 1912 says that:

> Three weeks ago there set out from Fremantle four whalers and a boiling down vessel on an experimental cruise of the ocean in the vicinity of Flinders Bay. On Saturday the *Fynd*, under Captain Bull, returned to Fremantle to replenish her bunkers. Captain Bull, who is in charge of the whole whaling fleet now in Frenchman's Bay, spoke of the success of the venture. Up to the time of the *Fynd's* departure from the temporary headquarters of the fleet, some fifty-five sperm whales had been secured. The majority of the catch were small whales which yielded only a small amount of oil, but at the same time the presence of a number of larger whales of the same species suggests that the old occupation of whaling is likely to be revived. Captain Bull stated that it was the intention of his company to establish a permanent oil extraction station on the north-west coast, and the necessary appliances are expected to arrive in the state in February next. Captain Bull in the *Fynd* leaves shortly on a cruise of the north-west as far as Monte Bello Islands, to select a site for the foundation of a northern base of operations.

That report proves that sperm whales were still to be found in commercial quantities and had certainly not suffered to the same degree as the right whales.

It also demonstrates the mettle of the Norwegians. Here they were thousands of miles from home, on floating equipment worth thousands of pounds, investigating the possibility of setting up shore stations on an uninviting part of a foreign country.

The organization difficulties were enormous, and the cost of transporting shore station equipment from Norway could not have been covered or even contemplated merely on the catch of fifty-five sperm whales or the "presence of a number of large whales of the same species."

Old whalers would have been delighted with such a catch, but modern whaling was not based on the cheap labour of the lay system and required a much richer harvest.

The records give somewhat confused reports of the activities of the Norwegians after leaving Fremantle, but the following, taken from a report by the state's Chief Inspector of Fisheries, is probably the most reliable:

During June 1912, the fleet of the Western Australian Whaling Company departed for the north-west coast. At the offset considerable difficulty was experienced in securing suitable harbours. A short trial was given to the waters east of Dirk Hartog Island (Shark Bay), but as this locality proved unsuitable, the fleet was taken to Maud's Landing where active operations were commenced. Later in the season the fleet was augmented by the addition of two more 'whalers,' and up to and inclusive of the early part of September, 243 humpback whales, of which 200 were males and forty-three females, were taken. On the approach of the warmer weather and consequently the approach of the completion of the migratory period of the humpback whales, the fleet proceeded to Frenchman's Bay (Albany), and the coastal waters in that vicinity were exploited for sperm whales. Again some difficulty was experienced in locating the 'trek' and feeding grounds. On the whole, however, the operations may be considered as having been fairly successful, ninety-eight sperm whales, of which seventy-nine were males and twenty-eight females, being captured to the end of December. The total take of oil during the season amounted to 7,800 barrels, of an approximate value of £22,000. In addition to this, a small parcel of ambergris was obtained from one or two of the sperm whales, but I am not in a position to give the value of this. The value of the whaling fleet is estimated at £80,000 and during the season an average of 115 men were employed.

It would be an exaggeration to say that the Norwegians

discovered a bonanza on the Western Australian coast in 1912; the gross returns on the capital outlay were adequate but not sensational, but they had established the presence of humpback whales, which no doubt pleased them immensely.

Writing of these, the Chief Inspector of Fisheries in Western Australia said:

> ·It may be of interest to point out that at least five species of whales frequent the waters of this state and, apparently, in point of numbers, they may be taken in the following order—the humpback, sperm whale, fin-back, sulphur bottom, and the right whale. In the migratory period— say June to October—the first mentioned species occur in our waters and journey as far north or perhaps a little further than Broome, whilst the second species is rarely seen further north than Geographe Bay. The fin-back is considered of little commercial value, they generally—in our waters at least—being in such a poor condition that their capture and treatment do not appear to be. a sound proposition. The sulphur-bottom, although perhaps the largest animal frequenting our waters—it grows to a length of about 100 feet—does not appear to be plentiful, and the right whale—one of the whalebone whales, and consequently very valuable—is met with in small numbers only, and at rare intervals.

And so the Norwegians were following the same pattern of operations off the Western Australian coast as their predecessors had done on the eastern and southern coasts of the continent in the previous century; whaling off shore for whalebone whales (the humpback having taken the place of the right whale) and then when the migratory period had ended, seeking sperm whales in the southern waters of the western state.

There was, of course, a big difference in accuracy, and the boom of the harpoon gun had taken the place of whispered orders of the headsmen as they closed in.

Later in the 1912 season the fleet was augmented by two more whale chasers, and it wasn't long before rumblings were heard in Perth from those who feared that the humpback would follow in the path of the right whale. An exclusive licence to hunt for seven years did not at the time include any restrictions on the numbers or sexes to be killed, but evidently the Inspector of Fisheries was not unaware of the dangers. On 9 January 1914, he stated:

During the year, two additional exclusive licenses were granted to companies formed in Norway—the Fremantle Whaling Company and the Cape Leeuwin Whaling Company. Other applications have, on behalf of Norwegian companies, been received for licenses, one particularly to cover that portion of the coastal waters extending from Cape Lambert to the northern boundary of the state. Consideration was given this application, and although the department is anxious to encourage operations, it was pointed out that as the periodic migrations in a northerly direction of the humpback whale were, in all probability, associated with provision for the next generation, it was not considered advisable to recommend the granting of a license covering those particular waters until further data had been obtained. Subsequent enquiries go to confirm my belief that in certain parts of the waters north of Cape Lambert are situated breeding grounds, and as I am anxious to avoid as far as possible the destruction of the females, and to protect the baby whales or calves, do not consider it would be in the interests of the state to, at this juncture, grant a license covering those particular waters.

This early statement on conservation was the first of many to be made in the following years but for the time the Norwegians were free to go ahead, and yet another company was floated to investigate our eastern coast.

On 10 August 1912, the factory ship *Loch Tay* of the Australian Company of Monsen of Tonsberg (not connected with the three in Western Australia), arrived in Sydney while her two chasers, *Campbell* and *Sorrell*, were searching the seas off New Zealand and Tasmania. The search was unsuccessful and the company eventually applied for a licence to use Jervis Bay. Here they ran foul of a naval establishment, and local residents who complained of the foul smell, and after three months of unsuccessful fishing in 1912 and five months in 1913 the company closed down and moved to Western Australia where by 1914 things were, according to the Norwegians' company reports, much brighter. These said that:

The north-west coast of Western Australia is very sparingly supplied with harbours and fresh water; a good and safe harbour for the success of our operations, and fresh water being essential, we waited with considerable anxiety to hear if this had been found. A few months after the ending of last season, a small harbour, the existence of which was known and it was thought only fit for small crafts, had been sounded. As regards the position of this harbour nothing could be better, as the track of the whale passed close in shore. The harbour was also big enough for several factory ships to lay there. This harbour is protected, besides a small island, by coral reefs that at low tide project a few feet above water. This harbour is lying at Point Cloates, at $22\frac{1}{2}$ degrees south latitude. It has been surveyed and christened Norwegian Bay. . . . In the last week of June the track of the whales started in earnest, the whales passing close by Norwegian Bay, and they were so close to the coast that the boats had often to back out so as to avoid touching the bottom. The whales increased from week to week, and the ships had more to do than they could manage. The *Prince George*, from which we did not expect a bigger production than 500 barrels a week, brought it up to 1,000 for some weeks. . . . It also happened that the boats had to lay in because the factory ships could not treat the whales as fast as they could be obtained; as many as twenty-eight whales were brought in in one day. The whole catch for the season was 23,430 barrels of oil.

The perseverance and skill of the Norwegians had paid dividends in the 1914 season and they were forced to leave the area two weeks before the southern track of the whales back from their breeding grounds had begun. If they had had the ships and men available, it would have been a catch equal to any known grounds for a season on the Australian coast.

After the harvest, the *Vasco da Gama* loaded some of the *Prince George's* cargo and headed for home, leaving one chaser in the area to continue observations while the remainder of the fleet sailed for Albany and sperm whaling.

The Norwegians were loud in their praises of the northwest, and even the climate, although described as "oppressive" was "never found to stop us in our work."

The delight of the Norwegians can be understood and on the strength of their success they intended to: "erect tanks in the ships; it will also be necessary at an early date to purchase a tank steamer for transport for the three companies. The factory ships would then be able to discharge their cargo into this boat without leaving their field of operations before the season was over."

Even in 1914, however, competition between whaling companies was keen. The Norwegians whaling in the northwest could not expect their successes to go unnoticed, but with an exclusive licence they felt fairly safe and their annual report summed it up optimistically by saying:

Even if the expenses in Western Australia are heavy they will be more than made up by the richness of the season, especially when it is remembered that we are not depending on the humpback season only. When this season is over we go south where the spermacet whale is found, and where we can operate the whole year round with this whale. We have, therefore, come to an understanding with the two other companies in Western Australia to work together; by doing this it will be possible to keep the industry going at more or less pressure the whole year round. The great advantage this will be to our company will be easily seen. On account of being in sole possession we need not be afraid of the whale taking off through damaging opposition. We have, therefore, a good opportunity to work up an industry that no other whaling company has been able to do before.

Their optimism, however, was premature; the clouds of war were gathering in Europe, news of their success had spread across the world, and as foreigners operating in a country at war, they were to be involved in its domestic soul-searching and internal squabbles.

12

You fellows go in for wholesale slaughter

DURING THE LATTER PART OF 1914 ADVERTISEMENTS appeared in Western Australian papers for witnesses to give evidence before a Select Committee to enquire into "the whaling industry generally, the granting of licences, the obligations imposed on such licences, and the standing and operations of the companies to whom such licences have been granted." 1,877 questions were asked and answered, and although the data gained enabled the committee to make sound recommendations, the proceedings also brought to light the prevalent ignorance and misapprehensions concerning the habits and customs of humpback whales, which had been migrating to the Western Australian coast for hundreds of years. The examiners themselves had to start from scratch, and must have surely tried the patience of the more experienced Norwegians and Australians who gave evidence. One passage between examiner and witness ran:

"Have you had any experience in whaling?"
"No."
"Do you know anything about the habits or customs of whales?"

129

"No."

"You do not know whether they have feeding or breeding grounds?"

"No."

"Who asked you to give evidence?"

"I saw an advertisement in the paper and I thought I could give some information in regard to the track of the whales."

"In what months would shoals be met with?"

"From May to the latter end of June."

"That would be going north?"

"I could not say which way they would be working. We find them playing about. They turn very often in their track. You can only judge that by the way in which they blow." (The witness retired).

The rumpus began when Captain Andersen of the Australia Company, failing to secure a licence to fish in Jervis Bay, New South Wales, turned his attention to the west coast. He gave the following reason: "Because the Federal Government have claimed all the foreshores, including Jervis Bay, and we cannot get a permanent lease. We had to waste about £9,000 worth of offal that we had boiled down."

The New South Wales Government did little to encourage the Norwegians, although they were willing to sign agreements similar to those in force in Western Australia.

On arrival in the west, Captain Andersen made a voyage up the coast, saw large numbers of whales near the Abrolhos Islands, and decided the area would be ideal for a shore station. Unfortunately it came within the licence held by the Fremantle Company, but this company had not at the time built a shore station and the captain, through an agent, made claim for a forfeiture of the licence on the grounds that the agreement had not been fulfilled.

The licence for the Abrolhos Islands was not granted but one in lieu was given serious consideration for an area further to the north-west. This, however, had always been prohibited, on the assumption that it could have been a breeding ground of humpback whales, and the three licensed Western Australian companies objected in very strong terms.

The controversy reached fever pitch, filtered through from ministerial level to the floor of the house, and then to the Select Committee.

From parliamentary reports, however, it is clear that the enquiry had a double purpose; to settle the dispute over the licences and at the same time to gather as much information as possible on the habits and customs of whales on the coast.

The Government seized the opportunity to gain knowledge, which without the assistance of the Norwegians could only have been had at great expense over a period of years, with the use of a research vessel.

One question very much in everybody's mind was the old one concerning breeding grounds. Some whalers maintained that humpback whales dropped their calves at random. Captain Andersen, who had been whaling for six years, said in answer to a question: "The whales have no breeding grounds." He further maintained that "they could drop their young in the Indian Ocean, up north or anywhere" and that "the theory that whales had definite breeding grounds was certainly not put forward by a whaling man."

During the enquiry the committee continually returned to this point in an effort to determine two things; if there was in fact a breeding ground on the northwest coast; and if it was thrown open, would the whales become disturbed, change their migration pattern, and find another breeding ground?

Captain Gustav Brun Bull, Manager of the Spermacet Whaling Company of Albany, was one of the most valuable and patient witnesses. When asked to define the breeding grounds, he said: "When whales are travelling they generally go at an average speed of five or six miles an hour and sometimes more. When they travel they make for their destination, which is the breeding ground. They follow the coast. Some tracks come in on the east part and others on the western part and the whales join up further north and go up in one stream along the north-west coast. They are maintaining the same speed all the way. They go around the

Monte Bello Islands and eastward of the Dampier Archipelago. When they reach the Dampier Archipelago they
have got to their breeding ground. Some of them go further
up as far as Bedout Island. Very few go further north. I have
been up as far as Derby and have only seen very few whales,
although now and then one is seen. I should say that the
breeding grounds are between the 117th and the 119th
parallel.''

Later, when asked to define the effect of allowing fishing
on the breeding grounds, Captain Bull said: "In my
opinion it would affect the whole industry, because whales
go up to the tropics where their breeding grounds are for
breeding purposes. If they are disturbed there, there is a
probability of their disappearing altogether. If they are
disturbed in their most vital haunts, the breeding grounds,
they may go up along the east coast of Australia instead."

If Captain Bull's theory was correct, it meant that
migrating whales were willing to suffer the inconvenience
of being hunted and killed *en route* to the breeding grounds
as long as there was a refuge at the end of the migration.
But if this sanctuary was invaded the whole pattern of
migration would be adjusted.

Captain Bull backed up his statement by saying: "On the
west coast of Africa when we started operations there, we
always thought that the breeding ground was outside the
Congo River. During the last year there have been too many
companies there altogether. The result is that the whales
have practically taken off altogether. It would be hard to
say where they have gone."

As strange as this theory may seem, it has an equivalent in
the case of migrating birds whose destinations are desecrated
by the sprawl of capital cities or the reclamation of swamp
lands. If their sanctuary is destroyed they do not cease to
migrate, but find more suitable domiciles.

But as far as the enquiry was concerned it was the desired
answer, as the government had all along maintained that
these northern areas should be kept closed.

It is interesting to note that those who wanted to fish in
the breeding grounds area maintained that there was no such

Below decks on a whale-factory ship. In the top photograph, the boilers are on the
right and settling tanks on the left. In the bottom photograph, the engine-room crew
are servicing centrifuges which separate different grades of oil (*P. P. Hey*)

thing as a definite breeding ground, whereas the three companies already established maintained that it would ruin the industry. The conservationists won the day and the area remained closed.

While picking the brains of foreigners in order to gain knowledge of whales on their own coast, the government also learnt several other things; one was that the Norwegians had assessed the value of the whaling grounds and given them a definite commercial life. Captain Andersen said that he doubted if companies could "make a living after ten years," while Captain Bull gave the area an estimated twelve years of commercial life.

Mr August Stang also summed up the grounds as follows: "I have before me a whaling journal which states that in South Georgia in one year 7,000 whales of one species alone were taken, and it must be borne in mind that over there are found many other and more valuable species than we have here. The fishing has been going on there since 1904 or 1905 and the islands still maintain their unquestioned supremacy as the finest whaling fields anywhere. They have further this advantage over Western Australia from the companies' point of view, that beyond the payment of the actual licence fees there is no local expenditure. I do not suppose there is a company anywhere who have spent approximately the amount of money we have spent locally."

The Norwegians were perfectly frank in their revelations, and with a little mental arithmetic it was clear that if they stayed the full seven years, which they were entitled to do under the terms of their licence, it would leave very little commercial promise for those who followed.

From then on the committee of enquiry returned again and again to the numbers killed. This was estimated at between 1,800 and 2,000 at Point Cloates in the 1914 season. And here another discovery was made as Captain Andersen answered questions. The dialogue between examiner and witness ran thus:

> "Do the whalers make any distinction in the whales they kill?"

Whale-catchers in action off the Cheynes Beach whaling station, Western Australia (*Nor' West Whaling Co.*)

"They could not do that."

"Do they kill anything they see?"

"If they saw two or three whales together they would try and kill the lot."

"And if they saw fifty together they would try and kill the lot, irrespective of whether they were male or female?"

"Yes."

"And irrespective of whether they had calves or not?"

"It is an understood thing not to kill whales that have calves with them, if it is possible to kill others."

"Suppose it is not possible to get another one?"

"Then they will catch the mother."

"And what happens to the young whale?"

"If it is big enough it will look after itself."

"And if it is not, I suppose the sharks will get at it?"

"The sharks do not touch anything that is alive."

"What happens to the young calf if the whaler kills the mother?"

"No one knows."

"Do you not know that it dies?"

"I do not know."

"Does your experience not teach you that?"

"I think if the mother is shot two or three days after the calf is born the calf will die because it cannot get milk."

"It seems to me you fellows go in for wholesale slaughter?"

"I never knew a whaler who made any scientific study of the whale."

It must have been in turn irksome, humorous, and embarrassing for the Norwegians to publicize the whaling facts of life through people who, at the beginning of the enquiry, were not quite sure of the difference between a right whale and a sperm whale.

At times, tempers became a little frayed, especially when the examiners returned again and again to the point of whether cows heavy with calves should be attacked. At one point Mr Stang was asked to "forget that you have any interest in these whaling companies and consider that you are interested, as we are, in the preservation of the industry, and yet trying to give fair treatment to the men who are whaling."

The Norwegians, however, adhered to unwritten laws of conservation, and as Mr Stang commented, had "introduced into the articles of the shooters a prohibition that females travelling with calf must not be shot. That was with a view to preserving the species. They may kill one by accident, but they would have to show it was an accident."

The Norwegians were commercial whalers, they had weighed the possibilities of the field, and been given a seven year licence, and although they knew it had a limited life would have been very foolish to allow indiscriminate killing to shorten their span of activity. They were commercial whalers out to make a profit, but not indiscriminate butchers.

From the maze of evidence the committee decided that "the three whaling companies at present operating have endeavoured by all means in their power to comply with the terms of their licences, and also to carry on their operations in such a manner as not to destroy the industry. The companies have spent thousands of pounds in exploratory work preliminary to fishing, and have in one case erected a large shore station costing approximately between £15,000 and £20,000, whilst another company is now erecting a station which will probably cost anything between £30,000 and £40,000."

The report also added: "Since the committee has commenced its work there have been presented to it figures which show that there is a great future for this industry if carried on under proper conditions and with proper safeguards."

This was no doubt meant to include a closed breeding ground, but as the Norwegians had given the field an estimated ten years life and three of these years had already passed, it seemed a little optimistic.

However, the Norwegians continued whaling, and apart from enquiries from Trade Unions about the wages paid to imported contract workers, interest in their operations again waned.

This was understandable. By late 1915 the papers had stopped numbering their casualty lists and were publishing columns of names of the dead, wounded, and missing; the conscription issue was tearing the country apart, and Billy Hughes was doing his best to convince everyone that Australia's future lay in the battlefields of Europe.

The Norwegians were neutral, but the by-products of the whaling industry were finding their way to Britain for use in the manufacture of explosives, and they were allowed to continue; there were more serious issues at stake than the fate of humpback whales with calves on the northwest coast.

The shadows of war, however, began to affect the confidence of shareholders in Norway. Mr Stang, in giving evidence, said: "At the outbreak of war all whaling companies' shares, being of a speculative nature, fell heavily...."

Added to this, the 1915 season, due to rough weather, was not as successful as previous years, and the Fremantle Company decided to end its Australian operations. The other two companies continued into the 1916 season, but again their returns were down on previous years and they too ceased operations before 1917.

And so the Norwegian era on the Western Australian coast ended, not because the whales had been hunted out, but because of the war in Europe. While men were busily killing themselves the humpback was left to multiply, and it was during this time that the Western Australian Government let a golden opportunity slip through its fingers. The breeding grounds had been pin-pointed, the government had been supplied with accurate knowledge of the track and times of whale migration, and certainly by then had the wherewithal to build whale chasers and harpoon guns.

Immediately the war ended, and its course had been made clear by 1917, Australian chasers could have been operating from the shore stations established by the Norwegians, supplying oil to European markets and fertilizer to Australian farmers.

By building chasers and cutting their teeth first on the coast, they would also have been in an ideal position after the war to sail south into Antarctic waters. Trained men would have been available, and with the provision of a factory ship (possibly a converted freighter) they could have been the first to exploit southern seas.

But many years were to pass before anything was attempted on a national scale, and it was left to individuals to attempt a revival. This took place in 1921, when a group of Western Australians launched a small company.

The history of this company was outlined by the Honourable Sir Edward Wittenoom on 26 September 1928, as part of a speech against a bill proposed by the Western Australian government. He said, "I happened to be the chairman of directors of the original company. The North-West Whaling Company was formed in Perth entirely by local people and with local money. The project was so popular that all the shares were subscribed in two days without a single advertisement appearing in the newspapers. Unfortunately, sufficient money was not asked for at the time. It was one of the cases where the management was to a large extent composed of people who knew nothing about the business. The company was carried on with scant success. First of all it had to take up a position at Point Cloates, many miles away beyond Carnarvon, where no means of communication existed. The company also had to procure the necessary vessels at a cost of a good deal of money. It was also necessary to gain a good deal of experience. The company did carry on for two years but because of the difficulty of obtaining a competent manager and a skilful gunner—in a case like this it is the whole secret of success—it encountered difficulties. In addition, a severe willy-willy which occurred on the coast wrecked one of the vessels. Under these adverse circumstances the company lost £24,000 in two years. The case seemed hopeless at the time, and it was decided to go into liquidation, but fortunately among the shareholders there were some progressive and liberal-minded men, who agreed to find sufficient funds for the purpose of sending a representative to Norway. The result was that another small company in Norway called the Norwegian Bay Whaling Company, agreed to work in with this company, and agreed to take over for a certain time the licence this company had obtained.

The second company have been doing that with a fair amount of success. At first there were some difficulties, but latterly the company have been paying the original company a proportion of profits that helps to liquidate the £24,000 I have referred to. It is hoped that by the end of the year the amount may be entirely paid off, and then the original company will be in a position to work the business. Up to that time, however, not a single shareholder had received a penny in dividends. All the money was put in by people in Perth, and not a penny has been received back. Now the government brings down this drastic bill containing clauses which, I believe, is unnecessary at the present with only one company representing the whaling industry."

And so the struggling Australian company was forced to call on the assistance of the Norwegians, through the good offices of Mr Stang, who made the trip to Norway.

It was clear that success in whaling entailed much more than a knowledge of the track of the whales; another reason for a thorough investigation and a far-sighted and even subsidized Government project.

The bill mentioned by Sir Edward Wittenoom was one brought down on 15 August 1928: "A bill to regulate whaling and the business of the treatment of the carcasses of whales for the purpose of obtaining commercial products therefrom, and to impose a royalty on whales taken in or brought into Western Australian waters, and for other relative purposes."

The issue was hotly debated, the government seeking a royalty of £1 per whale caught, and the opposition seeing it as unnecessary and dangerous, saying that, "Unfortunately the immediate effect of the bill has been that the manager of the company, Captain Bull, who is regarded as a leading expert on practical whaling, has recommended his company to cease operations this year. We must not do anything to drive the industry off the coast. If we pass hampering legislation the company may employ factory ships such as are employed in the Ross Sea that can handle the whales outside the three-mile limit and thus the state would run the risk of losing the whale industry. We propose to tax something that is not a product of this country or of any other country.

The whale is a migrating mammal that is the property of all, and it will be safe to say that not ten per cent of the whales caught by the company operating on our coast are caught in our territorial waters. Immediately individuals display enterprise, foresight and brains in the establishment of an industry where others have failed, governments proceed at once to impose taxation."

The opposition questioned the legal right of the government to impose a tax on migrating creatures, and one member facetiously suggested that all humpbacks passing up the coast should be branded.

While the bill was being debated, the Norwegians were in an unenviable position. They had been invited out to Australia, and their skill was gradually putting the single whaling company on a profitable basis. But the extra estimated £1,000 a year, which they would have to pay in royalties if the bill was passed, could have meant the difference between profit and loss. They became embroiled in domestic issues and a government policy which seemed determined to place obstacles in the way of a successful whaling industry, either foreign or Australian. As one opposition member suggested, the government should have been debating how much money to put into the industry, not how much it could make from it.

The Norwegians had proved that, given a fishing season with moderate weather, whaling could be profitable on the Western Australian coast; they had been honest in their assessment of the life of the field and had, for their own benefit and the benefit of the whale population generally, practised some attempt at conservation; they had also imparted their knowledge of whale customs and habits and in fact taught the Australians all they knew about modern whaling.

Much of it, however, had fallen on deaf ears, and by 1928 the very field which was being discussed had ceased to be important to the Norwegians. By that time, they had built and tested factory ships with slipways in the stern. These allowed the whales to be dragged straight up out of the water for processing, and did away with any need for shore factories. And they were so big that they could act as "mother ships" for the fleets of whale chasers, carrying all their needs and making them independent of shore bases. The whalers could seek their prey on the high seas, especially in the immensely rich Antarctic waters.

Migration patterns and customs were no longer important and before the bill had been defeated, by seventeen votes against and eight for, on the 26 September 1928, the Norwegians had decided to leave the Western Australians to work out their own problems.

Their production of 3,608,348 barrels (each of 375 lb) of oil in the 1930/31 season, the largest amount of oil ever produced in the Antarctic, made the Western Australian grounds insignificant as far as the Norwegians were concerned.

13

Australia should lose no time

"The master, captain seizo kobayashi, stated: 'we had a crew of 1.63 aboard the mother ship but did not have nearly enough to handle our catch. On our next trip we will be using another eighty men. All will be Japanese. We have learnt enough from the Norwegians to make a voyage with only our own countrymen'."

The above quote, which appeared in a letter to the *Sydney Morning Herald* on 22 March 1935, was used by a writer to strengthen his argument for an Australian whaling industry. He could not understand why Australians, "who could work the same areas with only a fraction of the overhead costs, have not sufficient business acumen to exploit such a source of wealth."

He was speaking of the Japanese success in Antarctic waters, aided by Norwegian gunners who sold their services and experience for a considerable salary to nations in competition with their own country's fleets.

Others were also pressing the Australian Government to establish an Australian whaling industry in the immediate pre-war years. On 17 October 1936, the famous Antarctic explorer Sir Douglas Mawson was reported by the Adelaide *Advertiser* as saying: "Australia should lose no time in establishing its own whaling industry." Even the Prime Minister, Mr Lyons, had said that "the Federal Ministry was convinced that if Australian companies were floated to exploit wealthy whaling grounds in Australian waters they would be an attractive medium for the investment of capital."

These statements came after the period in which whaling in Australian waters had gone into another steep decline, after the Norwegians had left in 1929. They had returned during 1936, for a brief burst of activity in which two fleets, operating under the American flag, killed over 3,000 humpbacks near Shark Bay, and again in 1937, when the Western Australian Government allowed operations in its waters with observers aboard. But compared with what was happening in Antarctic waters these operations were of no great significance, and certainly of no credit to Australians.

Perhaps the most promising move during these years, as far as Australians were concerned, was a journey made to England by Sir Macpherson Robertson, the man who sponsored the Victorian Centenary Air Race and financed the Mawson expedition to the Antarctic. On 22 February 1935, he defined his mission by saying: "The possibilities of founding an Australian whaling industry in the Antarctic with its headquarters in Tasmania, will be investigated. There is not sufficient capital available in Australia for the scheme, so I shall seek the support of English interests. For years the Norwegians, who have to travel for months to get from Norway to the whaling seas in the Antarctic, have been whaling profitably, but Australia and England have neglected an industry which is literally at the back door of the Commonwealth."

These statements brought an immediate response from Sir Douglas Mawson, who pointed out that: "There has been a whaling company in Sydney awaiting more financial support since 1929."

Australian investors were very cautious at this time, and unlike their predecessors of a century before, were not willing to back the industry, despite the fact that Norwegian and English interests were killing over 34,000 whales a year between them.

But their caution is easily understood. In one breath the experts were urging them to part with their money, but in the next were stressing the need for conservation and highlighting the dangers of indiscriminate killing.

In December 1938 Sir Douglas Mawson was reported as saying: "I still think that one of Australia's most glaring oversights is that it has not come into the whaling business. It seemed that because few people in Australia knew much about whaling they had left others to explore the industry." But later in the same statement he said: "It was believed to be a common thing for mother whales with suckling young to be killed and the young whales left to die. Whalers would give the excuse that they had not noticed the young whales until after the mother was shot. A heavy toll was being taken of whales today, and the only hope of maintaining them was not to kill the mothers with suckling young."

Two years before this, Professor W. J. Dakin, Professor of Zoology at the University of Sydney, had issued a similar warning, maintaining that "conservation was a difficult and vexatious problem."

But Sir Douglas Mawson was right. There were only a few people in Australia who knew very much about whaling, and they spoke with the frankness of the expert. No doubt their statements on the need for conservation, although they may not have intended them to do so, contributed to the caution felt by prospective investors at that time.

Another drawback was the huge outlay required for building and equipping a factory ship and a fleet of chasers, which would certainly have eaten into the profits for several years. Not to mention the lesser problem of obtaining trained gunners and technicians.

By 1938, however, the Commonwealth Government had come into the picture, and on 28 April the following appeared in the *Argus*: "The Prime Minister (Mr Lyons) today informed the Leader of the Opposition (Mr Curtin) that the project to establish a whaling industry in Australia had been revived, and that certain representations for Federal assistance will shortly receive consideration. Mr Lyons said that in 1936 a company owning plant and machinery at a whaling station in Western Australia had asked for financial assistance in establishing the industry. At that time aid was refused.''

Both government and private enterprise were confronted with a tantalizing problem. Their ports were nearer to the hunting grounds than the ports of any whaling country, and there was the added advantage that, as Sir Douglas Mawson pointed out, ''the industry could operate on the Australian coast in the winter and in the Antarctic in the summer, so that its activities could be in progress all the year round.''

But no one knew for how long. It was one thing to hunt whales as they migrated north to their breeding grounds, but now with deadly weapons they were being hunted in their feeding grounds. When disturbed in their breeding grounds, whales had been known to find new ones. So what would be the result of the invasion of their primary sanctuary? No one knew, and although some efforts were made to investigate the possibilities of an Australian industry, very little was accomplished in the pre-war years.

As far as conservation was concerned, however, there was a glimmer of hope as whaling nations began paying lip service at least to the idea of controlled killing.

During the seasons from 1932 to 1937 catches and production were controlled, either by private agreements between the whaling companies or by agreements entered into between the Norwegian and British Governments. Then from 1937 to 1940 it was controlled by the terms of an international agreement, signed by all countries engaged in pelagic whaling with the exception of Japan, which started Antarctic operations in the 1934/35 season.

By August 1936, Australia found herself in the rather ''also ran'' position of having to bring down legislation to control whaling operations in southern waters, although it did not own a fleet. ''The regulations provide that a ship, in respect of which a licence is granted, shall on each occasion on which the ship is despatched from Australia for the purpose of whaling, call at Hobart or at such other port approved by the Minister and must report to the Deputy Director of Navigation and Lighthouses in that port and produce to him their licence. It is provided also that in the case of a ship or factory there shall not be delivered to it a greater number of whales than can be treated by the plant therein, and that the plant must be capable of converting whales into commercial produce within 48 hours of the delivery to the ship or factory. The provisions set out in the Commonwealth Whaling Act are incorporated in the International Convention drawn up under the auspices of the League of Nations in 1931 and that convention has been ratified by 25 countries."

As a signatory Australia did her duty, but the period between 1929 and 1939 can only be described as one of lost opportunities. Perhaps this is not hard to understand, for it was the period of the Great Depression, when money was often hard to find for even the most basic needs. Though it might also be argued that a courageous government would have seized the opportunity to re-establish the industry while shipyards were begging to build ships at any price. And there may be another reason for the lack of public and government interest; that this was a period during which, most strangely for an island nation, Australia seemed to have little interest in ships and the sea, and was content to leave most maritime occupations to other countries.

Admittedly there would have been difficulties in entering the industry between the two wars, and it certainly would have tested the courage of private investors or alternatively the government, but after some attempt had been made for international conservation the future outlook seemed brighter.

The most prudent method would have been a government subsidy to allow the purchase of one or two chasers to operate from Australian shore stations (which had been established by the Norwegians and were awaiting renovation), and then, when the industry had been placed on a profitable basis, a factory ship could have been added to sail south.

Japan found it well within her capabilities, and no doubt

the Norwegians would have been willing to teach the Australians the lost art of whaling.

But in 1939 the war clouds again blew up over Europe, and whales received another respite while men attended to more urgent matters.

Immediately after the war, however, whales quickly returned to their former important position, because of a world shortage of edible fats. In November 1945, Mr A. R. Harrison, representing the British Ministry of Food, was reported as saying: "It is an interesting fact that in the surveys that we take of what the housewife most wants back of pre-war supplies, fats is the first thing." Later, speaking of the virtues of whale-oil, he said: "During the last ten years its virtues have been fully recognized and its few technical limitations determined, enabling whale-oil to regain a key position which it once held for centuries after having fallen almost into oblivion."

Similar compliments were being paid to whale-oil all over the world, and the stage was set for the greatest exploitation of whale stocks in the history of the industry . . . and for a hitherto undreamed of assemblage of men, machines, and scientific equipment to deliver a massive onslaught on the whale population. For six years of war, the whales had been left in peace while factory ships were used to carry fuel oil and gasoline to supply the Allied war machine, and whale chasers were used as convoy escorts. Now, a world which had been too busy fighting to feed itself looked greedily towards the replenished herds of the southern oceans.

In the first whaling era it was man against whale; in the second, the harpoon gun and factory ship against the whale; but in the third, men had equipment which revived the worst fears of the conservationists.

In 1946 the British Board of Admiralty decided to offer the older type of Asdic set, which was used by navy ships in the detection of submerged submarines, to whaling firms and to explain its uses. "A whale is seen spouting and the whaler closes the position. When the ship has closed to a certain range (500 to 1,000 yards) the whale probably becomes disturbed by the sound of the propellers and dives at speed.

In order that the whaler may be near the position when the whale re-surfaces, contact may be retained with the whale when submerged, by Asdic, accurate ranges and bearings of the whale being obtained. During this period, until the whale has become used to the noise and presence of the ship, it will be travelling at some speed, and there will be many occasions on which Asdic contact will be lost. On the other hand there will be other occasions on which the course of the whale can be determined and the overall chance of improving the number of kills may render the installation of Asdic sets in whalers desirable."

The Norwegians meanwhile were rebuilding their fleets with factory ships to accommodate 300 men, and also with the latest technical equipment.

Compared with these developments, the circumstances under which an Australian company caught its first post-war whale were pathetic, but the event was excitedly reported in a style reminiscent of that used to describe the industry a century before: "The crew of the Albany Whaling Company's chaser *Wadjemup* has made its first kill, a forty-ton humpback whale. Harpooned yesterday afternoon, the whale was captured after an exciting all-night tow off the entrance to King George Sound. It was the first whale killed by an all-Australian crew in Australian waters for many years. Whales were sighted off Cape Vancouver and at the entrance to the Sound one whale was harpooned twice, but its vital organs were missed. Darkness prevented further firing. The whale dragged the ten-ton chaser close to the rocks on which big swells were breaking and for some hours the boat was in danger. Later a strong wind sprang up and the superstructure of the boat acted as a sail and gradually pulled the boat from immediate danger. At dawn the whale was still full of fight but was obviously badly hurt. It was not until after 10.25 a.m, however, when two more explosive harpoons were fired that its display of fortitude ended. The struggle lasted seventeen hours."

At a time when the latest advances were being utilized in the Antarctic, and thousands of whales killed, an Australian private whaling company was using a ten-ton chaser, and taking seventeen hours to despatch a single whale.

The next private effort, that of the Australian Nor'-West Whaling Company at Point Cloates which began in July 1949, was better equipped but still left a lot to be desired when compared with overseas firms.

This firm took over the old whaling station established by the Norwegians, and after renovations the equipment included two steam boilers, fifteen digesters and a forty-ton whale hauling winch. The whales were flensed on a platform extended from the water's edge to the bottom of a ramp leading to the cutting up deck above the digesters. As the blubber and meat was stripped from the whale it was hauled on to the ramp by a winch and cut up into small pieces and fed into the digesters. The average time for a cook was six hours and even with this rather primitive equipment, about six tons of oil per whale was obtained.

The average price for oil at that time was £70 sterling which meant an approximate return of £525 Australian per whale. In the first season most of this was sold to the Dutch Government. The company also built a meal mill which produced about sixty tons of protein meal, which along with guano made from the offal, was sold locally.

The outlay was approximately £100,000 and the station employed 105 Australians living in barracks and tents. But the two key personnel, the gunners, were Norwegians, Captain Larsen and Captain Jenns Anderson.

The company's main chaser, the *Norwegian Bay*, was a converted Fairmile naval coastal patrol boat, with a twenty-five foot mast which could be folded down to pass under Swan River bridges. A tractor was mounted forward, with its power drive operating rope-hauling drums instead of caterpillar tracks.

This company demonstrated, in two successful seasons on a limited capital, what could be done on the Australian coast. With the Albany Whaling Company, they were the pioneers of post-war private whaling. Despite the loss of one chaser in a storm, the Nor'-West Company survived long enough to figure in the débâcle which followed the Australian Government's bid to enter the industry.

Point Cloates whaling station, Western Australia, which functioned from 1912 to 1920 (*Library Board of Western Australia*)

By this time, whale-oil had become a very valuable commodity, and Norwegian papers were reporting that oil produced in the 1949-50 season had been sold in advance for £80 a ton.

Other · sources listed seventeen whaling fleets in the Antarctic in the 1948-49 season producing oil worth over £33 million (Australian). These comprised ten Norwegian, three British, two Japanese, one South African, and one Dutch.

Some control methods had now been realized through the International Whaling Convention signed in Washington in 1946 by eighteen nations, including countries engaged in pelagic whaling in the Antarctic and an Australian representative.

Many whaling interests, however, only saw it as paper law and not binding on whaling nations. In the *Fisheries Newsletter*, February 1963, D. J. Gates of the Fisheries Division of Primary Industry, wrote:

> The main reason for the failure of the Commission, which administers the Convention, to maintain the whale stocks at post-war levels is found in the Convention itself. Here I refer to paragraph three of article five, which permits each contracting government to object within ninety days to amendments relating to conservation measures as made by the Commission. Once a government objects to a particular measure, this measure is not binding on it and so the measure becomes ineffective. An objection by one government usually results in other members lodging similar objections, and so in a regulation passed by the Commission not being observed. The fact that the Commission has no powers by which it might enforce its regulations is another reason why it cannot achieve what it set out to accomplish.

Probably the main cause of the decline of the whale stocks has been the method used to limit the Antarctic whale catch. Except during the years 1949 to 1953, when the humpback catch was limited by numbers, there was no move until 1960 to restrict the catch of each species on a

The whale-catcher *H. J. Bull*, seen here during the 1937 whaling season in Shark Bay, was the largest and fastest of her class in the world at that time (*Library Board of Western Australia*)

Whales secured astern of the whale-factory ship *Ulysses*, during the 1937 whaling season (*Library Board of Western Australia*)

population basis. As a result the fin whale populations, being much greater numerically than other species, have supported the Antarctic pelagic whaling industry and been greatly reduced. At the same time, the blue and humpback whale populations have been reduced to the verge of extinction.

Australia was vitally interested in the numbers of whales killed in Antarctic waters in post-war years, particularly as far as the humpback was concerned, for as previously mentioned, only two of the five almost self-contained humpback populations migrated to Australian waters. Any reduction in the species as a whole in the feeding grounds would therefore affect the numbers of the two Australian migrating populations. If at any time Australia decided to venture into the industry from its own shores, it would have to satisfy itself that there were still sufficient numbers on which to base an industry.

14

The people's assets

"Fishermen of western australia are asked to regard as important and urgent the request made to report to the Commonwealth Director of Fisheries the appearance of any whales they may sight. If a sufficient number of reports are received from observers, the information so obtained will be of real value in helping to plan the establishment of a whaling industry in Australia on sound and lasting lines—a cause which every Australian fisherman will naturally have at heart." (*Fisheries Newsletter*, August 1948.)

By the time an Australian Government was willing to do something about establishing a whaling industry, the information gathered and made available by the Norwegians between the wars was obsolete, and that of the few Australian private companies which had sallied forth in post-war years was inadequate. Whale stocks had been affected by intense Antarctic whaling, and the above plea to Australian fishermen was sent out.

The information desired was: "Date when whales sighted, locality, number of whales, species, apparent size, whether with or without calves, direction in which whales were travelling, and the fisherman's name and address."

The Director of Fisheries offered to acknowledge every report and refund the postage and the response was described as "splendid." Reports indicated that whales were numerous on the Queensland coast at Lady Elliott Island, Double Island, Stradbrooke Island, and Cape Capricorn, from July until the end of September, when they commenced their southward trek. In the west the northerly trek past Rottnest Island continued from May until August.

By waiting until 1949, however, the Chifley Labor Government had committed itself to far greater expense than that of return postage.

The *Fisheries Newsletter* editorial of April 1948 summed it up in this way:

> It is time Australia made some effective attempt to utilize the whaling resources of her coasts. In fact, there is no time to lose, for with whale-oil in such urgent demand and so highly priced, it is only to be expected that if we delay much longer, somebody else will step in and help themselves to this undeveloped wealth. This would be all the more regrettable because instead of importing 1,500 tons of oil a year, there is no reason why we should not supply our needs and still have a large surplus for export to hard currency and dollar areas. But whaling must be undertaken at the highest technical level, using the latest methods and most modern equipment, and taking advantage of all the experience acquired by established whaling companies of Norway and Great Britain. To enter upon whaling in any piecemeal fashion, or with makeshift equipment, would merely be courting failure. To assist in establishing an Australian whaling industry on a sound basis, the Commonwealth Government has brought to Australia one of the world's leading experts in Capt. A. Melsom.

Captain Melsom had been whaling since he was eighteen. He made his first trip to Spitzbergen in 1907, and by 1913 was a gunner-captain. He visited Point Cloates in 1914, and between 1919-29 was whaling for a Japanese firm, Toyo Hogei Kaisha, in Japanese waters and in the Antarctic.

He was reported to have killed some 3,850 whales during this time.

His services were an added expense to the plan, but whaling had undergone such radical changes that the advice of such an expert was essential.

He was engaged for twelve months as an adviser, and in company with the Director of Fisheries inspected all possible whaling sites on the Australian coast. His advice to Australia was to purchase a factory ship and enter Antarctic whaling, or, if this was not possible, to set up a shore station near the normal track of the whales.

Overseas inquiries revealed that a factory ship would cost between £2 and £3 million pounds. The government recoiled at this, and while making further inquiries decided to begin shore-based whaling. Two officials were sent to Norway to purchase equipment and to engage experienced personnel, and in May 1949 the Commerce Minister, Mr Pollard, introduced the Whaling Industry Bill to authorize the establishment of the first publicly owned shore station in Australia.

The Bill had a surprisingly smooth passage, receiving very little real opposition from those who were party bound to resist anything which could be described as socialistic.

The governing body, known as the Australian Whaling Commission, had its headquarters in Melbourne, and an office in Perth to carry out administration procedure. Babbage Island, at Carnarvon in Western Australia, was chosen as its operating site. This was a wise choice; it had a good supply of fresh water, was connected by road to Perth, and was a regular port of call for coastal ships. The three whale chasers, *Carnarvon* of 600 tons, *Gascoyne* of 300 tons and the *Minilya* of 250 tons, were modern and efficient.

The A.W.C. also imported fifty-nine experienced Norwegians under contract to put the business on a profitable basis and to train Australian workers, and by 1952 it was employing 162 men on the catchers and on shore, and twenty-three people on administration.

The project was financed by the Treasury for just over £1 million, with provisions for extra working capital, if necessary, from the Commonwealth Bank's Rural Credit Department.

The processing plant was capable of handling 1,000 whales per season and was equal to any in the world. Working conditions were fixed by industrial agreements negotiated at the beginning of each season; wages being fixed partly on hours worked and partly by bonus payments—which, with their resemblance to the "lays" by which they were paid, were about the only part of the entire scheme which would have been recognized by the pioneer Australian whalemen who challenged the whale from their flimsy boats. During the seven years of its existence, the A.W.C. did not lose an hour's work over industrial disputes.

And so, at long last, Australia was in a position, with modern equipment and experienced men, to exploit the wealth which lay at its doorstep.

If it had done this at any time between the two world wars its task would have been much easier, but by 1949, even on its own coast it was fishing in competition with all and sundry and under the terms of the International Whaling Agreement was restricted to a quota for each season.

Its processing plant could handle 1,000 whales a season, but in 1951 its share of the Australian quota was 650 humpbacks, for the next three seasons 600, and for 1955 it was 500. One catcher was kept in reserve, and the A.W.C. usually caught its limit before the season ended.

Meanwhile the Nor'-West Whaling Company was still operating at Point Cloates, and between them they took 1,224 whales in the 1951 season from which over 9,000 tons of oil was produced. This proved that even limited fishing could be profitable, and strengthened the arguments of those who maintained that much bolder plans were needed. The *Sydney Morning Herald* said:

The work proposed by the Commission is only a beginning. The plan approved by the Chifley Government envisages off-shore whaling based on coastal stations. While this should be useful and profitable, it is no substitute, from the viewpoint of rewards, for the development of our vast whaling assets in the Antarctic. Australia lays claim to no less than 2½ million square miles of the richest whaling area in the world and is doing nothing about it. Lack of a factory ship has been the excuse offered, but it is open to doubt whether any very determined efforts have been made to secure one. The new minister responsible, Mr McEwen, should make an early investigation of the whole whaling problem with a view to beginning pelagic operations at least by next season.

The cost of a factory ship had always been a stumbling block, but as early as November 1948, Prime Minister Chifley had made enquiries and even went as far as seeking one from the Japanese as part of war reparations, but nothing eventuated.

Whether his effort could be described as "determined" or not, the fact was that by the time the A.W.C. was operating successfully the old bogey of over-fishing was receiving even greater publicity, and no doubt tempered his efforts in this direction. Australia had missed the boat as far as Antarctic whaling was concerned in more ways than one.

Meanwhile the government (now Liberal) had gone as far as putting whale steaks on the menu of the Parliament House dining-room at Canberra, in October 1950, and the A.W.C. shore station seemed destined for several years' profitable life.

Up to 1952 the east coast of Australia remained virtually untapped, but in that year a third company, Whale Industries Ltd, was registered as a public company in Sydney, with a nominal capital of £1,000,000 in 4,000,000 ordinary shares of 5/- each.

By this time, having witnessed the success of the A.W.C. and the Nor'-West Company, Australian investors were now confident.

The birth of the new company, and its policy, had much in common with its predecessors on the west coast. "After careful investigation the company has decided to operate from a shore based treatment station during the May to October migrating period. In making this decision the company took into consideration the high capital cost (approximately £3 million) and operating cost of a factory ship.''

The company also fell into line with the government, and paid a compliment to Captain Melsom, by appointing him the company's whaling master and giving him a seat on the board. Under his guidance they chose a thirty acre base on Moreton Island, near Brisbane, to erect a treatment plant, close to the track of the whales, and near enough to a capital city to ensure the economic supply of fuel and other requirements.

In the same year, the Cheynes Beach Whaling Company commenced exploratory sperm operations from Albany in Western Australia, and in 1954 a fifth station was established at Byron Bay, New South Wales, and a sixth at Norfolk Island in 1956. These last two were operated by the Norfolk Island and Byron Bay Whaling Company.

Each of these companies had their own story to tell. In less than seven months Whale Industries Ltd erected all its buildings and a plant designed to treat nine whales in twenty-four hours. Steam was provided by an oil-fired boiler and electric power from a steam driven generator, and a system of pipelines to carry fresh water to the catchers, and salt water to the factory, was installed.

Meanwhile at Byron Bay in New South Wales, whales were hauled up the slipway on to a rail flat top which took them to the factory about 600 yards from the jetty.

Although operating on limited budgets, and at times having to improvise their equipment, the private companies, along with the A.W.C, reached a high peak of efficiency. In 1954, a total of 2,001 whales was taken. This netted approximately 102,000 barrels of whale-oil, all of which was exported, 14,750 tons of whale meal, and 1,809 tons of solubles. Together they brought over £1¾ million, and by 1955 this had jumped to £1,953,140.

During this time, the A.W.C, having greater facilities for research and technical improvements, had shared its knowledge and, at times, its quota with the private companies, and a happy situation had developed within the industry.

But still no moves were made for the provision of a factory ship, and Australia contented herself with limited shore fishing.

There had also been a change of administration, and the Hon. William McMahon had become Minister for Primary Industry in a Liberal Government. In the March 1956 issue of the *Fisheries Newsletter* he was reported as saying:

Even in a meat eating country like Australia, fish is needed for variety and for its special nutritional values, particularly in institutional and invalid diets. And of course there is the national need to increase exports to which our fisheries resources, if properly developed, could make an even greater contribution; for example, prawns. In face of this double need for increased fish production, the catch is unfortunately not keeping pace with our rapidly growing population. Normally, imports provide about half the supply of fish available for consumption in Australia. The last few years should have taught us how unwise it is to rely on imports for essential needs. The recent restriction of imports of course includes fish. Moreover, if we can produce more fish and thereby import less, or at least not an increasing quantity to meet the growing shortage, we shall correspondingly improve Australia's trade position. There is, therefore, urgent need to explore our latent fishing resources and to begin, as soon as possible, to develop them. Nobody, of course, would expect me already to have come to any conclusion as to what might be the best way of going about fishery development in Australia . . . . It is my desire as Minister for Primary Industry to see fish production greatly increased, both for domestic consumption and also to provide exports to help pay for the imports which Australia must obtain for its national development.

In March 1956 the Minister may have been undecided, but by May the same year, fishing journals and newspapers were headlining: "Big fund to develop Fisheries from sale of A.W.C to Nor'-West Whaling."

In his second reading speech on the Fishing Industry Bill,

establishing the Fisheries Development Trust Account, the Minister for Trade, Mr G. McEwen said:

The purpose of this Bill is to establish a Fisheries Development Trust Account which will be used to foster the development of the Australian fishing industry. It is intended to finance the Trust Account from the surplus which will arise from the sale of the Australian Whaling Commission's business at Carnarvon in Western Australia. With the sale of the Commission's business, the Government will have recouped its total investment in the whaling industry and will at the same time have a very substantial amount in hand. Normally this would be paid into Consolidated Revenue. However, the government in its determination to develop Australia's resources to the full, to give a broader base to the economy, to earn overseas exchange, to reduce expenditure on imports and to introduce improved techniques, now proposes to develop other fisheries enterprises. In short, the profits which have been made from the government's whaling operations will be isolated in a special Trust Account to assist in a special way the development of the fishing industry. So we have the present spectacle of having completed a demonstration of enduring value to the whaling industry at no cost to the taxpayer and now, with the profits from this demonstration, of endeavouring to repeat this performance in other sections of the fishing industry.

The government was right in its assessment of the A.W.C; it had certainly served a purpose apart from making profits, and was a show place to which came many visitors, both Australian and overseas. It had tested various types of catchers and processing equipment to discover those best suited to Australian conditions and had trained Australian workers, and as far as the government was concerned, it had served its purpose. Its attitude was the same as that towards the sale of the public interest in Amalgamated Wireless and the Commonwealth Oil Refineries.
The Minister for Trade, Mr McEwen, said: "The con-

tinuance by the government of purely productive enterprise could be justified only by adherence to socialist doctrines."

The *Fisheries Newsletter* of May 1956 also backed the scheme, saying that: "The sale will make possible the biggest move in Australia's history for the development of the nation's fishing resources, for the surplus from it, which may amount to £750,000, will be paid into a new Fisheries Development Trust Account for that purpose. Professional fishermen throughout Australia will welcome this good news."

At first sight the scheme did have some merit. The undeveloped fishing industry certainly needed a boost, and a windfall of £¾ million would allow for research and development. By selling the A.W.C to another operating company Australian exports would not be affected as the licence would be transferred automatically and the purchaser could double its production. This would mean extra staff and afford some protection for the employees of the A.W.C.

But a closer examination of the A.W.C records would seem to suggest an alternative. R. L. Wettenhall, writing in the *Australian Quarterly* of December 1961, summed it up when he said, "The project was financed by Treasury Advances totalling £1,375,000, and although arrangements were made for additional capital to be obtained from the Commonwealth Bank's Rural Credits Department, this figure proved more than ample. The establishment had been completed in time for the 1951 season at a cost of £1,080,840, and profitable operations, thereafter, enabled the A.W.C to cover its own insurance, write off substantial amounts for depreciation (by 1956 the capital cost had been written down to £809,000), pay all taxes other than income tax (from which it was exempted), and repay £850,000 of the Treasury advance as well as interest amounting to £126,025."

And so the Government appeared to be quitting a regular source of income in exchange for a lump sum. There were those who believed it would have been wiser to retain the A.W.C and sink its profits into the fishing industry year by year. This, of course, would have meant a slight alteration to the Liberal policy of not operating in competition with private enterprise. But they managed to find it in their hearts to do this with Trans Australia Airlines, and no doubt could have managed the same with the A.W.C.

If they had chosen this course, all things being equal, Australia could have had not only a thriving fishing industry but perhaps a factory ship operating in the Antarctic at the present time.

But it must be said that, to have made a success of whaling over a long term, the purchase of a factory ship with a view to entering Antarctic whaling industry was the only answer. By 1958, the writing was on the wall for shore-based whaling. Whether the government realized it or not, subsequent events were to prove that they sold the A.W.C at the right time.

The opposition, however, saw things in a different light. When the bill to set up the A.W.C was introduced it had a mild passage through parliament but the one to dispose of it had a stormy one.

On 3 May 1956, Mr Pollard said: "It is now fairly obvious that we have reached a stage at which, so far as the House of Representatives is concerned, the assets of the Whaling Commission will be sold. But there is always the reservation that the measure has yet to run the gauntlet of the Senate. I hope, and it is possible, of course, that some honourable senators from the government ranks may see fit to oppose this obnoxious bill."

Later in the same debate Mr Pollard remarked: "I hope that, before very long, there will arise circumstances that will give the present opposition the opportunity to exercise constitutional rights in order to restore to the Australian people the public assets that are being disposed of. In addition, I wish to point out that notwithstanding the parade of virtues by government supporters who claim that they believe in competition, the sale of this instrumentality means that, for the first time in the history of the rejuvenation of whaling in Australia, one company will hold a licence to take at least 1,000 whales annually. Normally it has been the practice to restrict one undertaking, private or government, to a quota of 500 whales. In future, one company will process at one factory, which has the necessary capacity, 1,000 whales annually, and the element of competition will be removed. Of course, that is in line with the policy of the government parties. They support in every way all these private enterprises and instrumentalities in the community which tend to close all avenues for an individual to become the proprietor of a small business. That is their policy. . . . One might ask why I say that this action will lead to the processing of 1,000 whales by one station. The apparently successful purchaser of the Carnarvon station is a company which at the moment is operating the Point Cloates station. The fact that this station ever operated is a tribute to the engineering ingenuity of the persons concerned. I have always appreciated that fact and I recognized it when the licence was issued. They rejuvenated a worn out, decrepit, broken down outfit, and it is well known that that station, after having operated for five or six years, has now reached the stage where most of its plant, machinery, and equipment is on its last legs, and they have been waiting like vultures for this instrumentality to fall into their hands in order to avoid the cost of restoring their outmoded and outworn plant, and to walk into the Carnarvon station, which is in first class order and condition, and is the most thoroughly equipped and efficient plant in the world.''

The opposition made much of the point that by selling to the Nor'-West Whaling Company the government was creating a monopoly. They also criticized aspects of the tendering procedure and accused the government of conducting it behind closed doors and of denying the Western Australian Government an opportunity of bidding. The sale price also came under attack as a figure far below the working value of the station.

And in its seventh and final annual report the A.W.C said that ''It was not taken into the confidence of the government when the decision was made to sell its assets so that the knowledge and experience of members could have been availed of to protect the interests of the Commonwealth when the conditions of sale were being proposed.''

The debates on the bill were fiery, and to those with no stake in the matter they were entertaining. Members were at their best, and even Moby Dick, the great elusive whale, was mentioned once or twice.

Opposition senators shouted and jeered when they were prevented from speaking, and there was an uproar when Senator Spooner, the Minister for National Development, said that the government would force the bill through if opposition members used delaying tactics. Later the Senate declared the Whaling Bill urgent, and applied the closure at all stages to curtail discussion, which was described by Senator Donald Grant (Lab. N.S.W) as "a world record in gagging."

However, the opposition was powerless, and the A.W.C was sold to the Nor'-West Whaling Company for £800,000.

The whole thing had been handled in the best parliamentary circus tradition by a determined government and a weaker opposition, strictly on party lines, and the only promising thing to emerge was a move by the government to develop the Australian fishing industry.

For those who saw the A.W.C as a modern efficient shore station (although established a decade too late), and the stepping stone to Antarctic whaling, it was a sad day. They realized that although private companies had made a success of shore based whaling, none was in a position, or ever likely to be, to enter Antarctic whaling. The old Australian whalers had left their mark in most known whaling grounds of the Pacific, and even further afield, but the new generation were to be denied the opportunity of working an Australian fleet in the south.

After the purchase of the A.W.C the Nor'-West Whaling Company closed its station at Point Cloates, and, as the opposition had prophesied, transferred its quota to the more modern plant at Carnarvon. There, in 1957, it processed 1,120 whales.

Meanwhile, at Albany, the Cheynes Beach Whaling Company had managed to erect a station which they improvised from old mining equipment and an unused wheat distillation plant from Collie. With this, they were successful in processing whales, and some of the ways in which this species can be used in modern living is outlined in a report from the company:

Sperm oil, which is more of a liquid wax is used as a lubricant for engines (because the viscosity changes very little in wide ranges of temperatures, making it valuable in fluid drive transmissions). Sperm oil is also used to impart a rich glossy sheen to cosmetics such as facial creams and it also gives a softness and flexibility to leathers. In textiles sperm oil lubricates the fibres to prevent unravelling as they are twisted into threads, and it is used in detergents, wetting agents, fibre softeners, hand soaps, and rust proofing compounds.

A far cry from lamp-oil and candles, and these new uses made it possible for the Cheynes Beach Whaling Company to continue while others were battling with dwindling humpback stocks and quotas. The company gives the following account of its treatment procedure:

After the whale is pulled up on the flensing deck, the whale's blubber, or outer layer of skin and fat is pulled off in long strips in the same way as you would peel a huge banana. The flesh is then cut away and the bones sawn into pieces. All these sections are slid through holes in the deck into the digesters, or boiling pots below. The great 'stew' of about forty tons of blubber, meat and bone is boiled under pressure for about three hours; when the cooking is complete the oil, as with the fat in a household stew, is at the top and is tapped off to be put through centrifugal separators in which the pure oil is separated from any remaining water and solids. The remainder of the 'stew,' that is the gravy and the solids, is passed through machines called 'super-d-canters' which separates one from the other, the solids going one way, the fluid the other. The fluids once again are put through a centrifugal separator in which the oil is taken away from the water. The solids in their turn are fed into a large oven where they are dried to become 'whale-meal.' This whale-meal is a dry powdery substance which is very rich in proteins and is used in the manufacture of stock and poultry foods. A whale produces nearly two tons of whale-meal. After the process has been completed, there is a large volume of dirty liquid left over known as 'gluewater' or 'stickwater,' and is further processed by putting this material through a three-stage evaporator which in turn produces a substance like that of treacle. This is then poured over hot steam-heated drum rollers which dries instantly into a thin skin and is scraped off the drums by sharp blades. This is then pulverized very fine through another machine, and is bagged off like fine cocoa. This product is known as 'whale-solubles' and is more rich in proteins than whale meal, but is used for the same purpose. More than two and a half tons of solubles is produced from each whale. The oil, the most valuable product of the whale, each mammal producing between six and seven tons, is shipped in bulk to United Kingdom.

While sperm whalers managed to make ends meet, humpback whaling went into a decline from 1960 onwards, caused by a reduction in quotas, and a reduction in the length of whales caught.

In February 1964, the *Fisheries Newsletter* printed the following by D. J. Gates, Project Officer, Fisheries Branch, Department of Primary Industry:

Estimated value of whale oil and by-products in 1963 was £512,000, compared with £1,006,000 in 1962. Following the collapse in 1962 of the Australian humpback whale fishing, three whaling stations, two on the east coast (Tangalooma and Byron Bay) and a third on Norfolk Island, closed down. The two remaining stations, located at Albany and Carnarvon, Western Australia, continued humpback whaling in 1963. These stations were allocated quotas of 100 and 450 humpback whales respectively, for the 1963 season. Catcher vessels operating from the two stations were also engaged in sperm whaling.

As anticipated, both companies experienced difficulty in obtaining a reasonable number of humpback whales,

The jaws of a sperm whale, with tongue removed (*P. P. Hey*)

Cutting-in a whale on deck. Note the feathery whalebone in this whale's jaw. In all species, except the sperm whale, this is used to strain out the tiny crustacea on which the whale live (*P. P. Hey*)

the total catch being only eighty-seven. The Cheynes Beach Whaling Company, which started sperm whaling in March, took its first humpback whale on 13 June and nineteen whales had been obtained when sperm whaling was resumed. Humpback whaling started at Carnarvon on 6 July, and only sixty-eight whales had been taken over a period of thirty-five days when the management terminated operations and planned to resume exploratory sperm whaling which it had started on 3 June. Two catcher vessels, supported by a spotter plane, were used in operations at Albany, whilst five vessels and a plane were used at Carnarvon. Only two vessels were in use during sperm whaling operations from the latter station. Sperm whaling, which the Cheynes Beach Whaling Company pioneered in 1955, is now the only commercial whale fishing in Australian waters. During 1963 a total of 574 sperm whales were taken at Albany and twenty-four were taken off Carnarvon during exploratory operations.

The switch to sperm whaling was hastened by a decision of the International Whaling Commission to prohibit, for an indefinite period, the capture of humpbacks south of the equator.

According to scientists it will take up to thirty years' complete protection to restore them on the east coast, and up to eighty years on the west coast; and even then these areas could only support whaling on a limited scale.

The humpbacks supported the industry for many years. They brought Norwegians to our coast, to supply, in the first instance, modern equipment and technical skill, and in later years provided the individuals who guided us through the re-establishment of a whaling industry. Without their assistance, the establishment of the A.W.C and the success of private companies would certainly have been considerably delayed.

But what of the future? It now depends, it would seem, as far as offshore whaling is concerned, on that old wandering, rugged individualist, the sperm. Through the pages of

Factory operations at the Nor' West Whaling Co., at Carnarvon, Western Australia. The whales are being hauled up the ramp for cutting-in and trying-out in the factory on shore (*Nor' West Whaling Co.*)

history he seems to have survived where others have been hunted almost to extinction.

In the old open boat days he had more than a fighting chance, and since he is mainly a wandering warmer water creature, he escaped intense slaughter. But now that humpback whaling has ceased off our coast, he is marked for immediate attention, and is being hunted with new equipment including spotter planes.

In April 1963, under a £24,000 grant, the CSIRO division of Fisheries and Oceanography began a two year survey of sperm whale resources. Using a twin-engined aircraft, flying over deep waters beyond the continental shelf, some 4,500 square miles of ocean were searched each month. Numbers sighted, size, and length were taken into account in an effort to assess the commercial life of the sperm whale.

Meanwhile, the exploitation continued. 710 were taken in 1964, 668 in 1965, 606 in 1966, and 587 in 1967.

Unfortunately, the sperm in Antarctic waters has also received more attention in recent years, because of the restrictions on killing other species. In the 1965-66 season, some 4,538 were taken. Although migratory patterns between the southern oceans and Australian waters are not as closely linked as with other species, the toll will certainly have an effect on the overall sperm population.

Great care will have to be taken in future years by all nations, including Australia, to make it possible for us to continue exploiting the sperm whale on which the industry was founded in the previous century.

In fact, the future of world whaling can be said to be in the balance. Blue whales and humpbacks are now protected in all seas, and the International Whaling Commission was forced to limit the pelagic catch for the 1966-67 season from 4,500 to 3,500 blue whale units (one blue whale unit equals two fin, two and a half humpback or six sei whales), and in 1967-68 from 3,500 to 3,200 units.

To many, the history of whaling has illuminated the best and worst characteristics of mankind. In the early days, it demonstrated his courage and endurance, as he matched his puny strength against the giants of the sea for a miserable wage. In the modern era, it has pinpointed his inhumanity and total disregard for the infliction of suffering, in hunting these splendid creatures almost to extinction, with small risk to the hunters and using equipment and weapons which give the whale no chance of escape.

Even women now have a hand in the killing. On Tuesday 21 May 1968, the 44,000 ton mother ship *Soviet Russia*, and twenty hunter ships, called at Sydney on their way home from killing 3,321 whales. On board were fifty women: stewardesses, nurses, doctors, meteorologists, and lecturers.

Conservationists maintain that whaling should be stopped immediately to preserve the species, while others would like it stopped purely on the grounds of cruelty to dumb animals.

Another group, hovering in between, see the Antarctic as a great food-producing area for a starving world and would like to see whales killed only for human sustenance. They point out that every by-product of the whaling industry can now be made from vegetable, mineral, or synthetic materials.

But when one considers man's weakness for profit, and his inability to frame workable international laws, in all probability whaling will go on under the present system until there are so few whales left that it is not worth setting out in pursuit of them.

And this appears to be the only hope for the survival of the remaining species; that, like the buffalo of America, the few survivors will be left in peace to build up their numbers once again.

Australia has always been conscious of the need for conservation, and has placed inspectors aboard its ships under the terms of international agreements. These measures, while affording protection in our local waters, were useless in controlling (in the case of the humpback) the exploitation in the Antarctic, whence the Australian whale population originated.

Without an Antarctic fleet of its own, Australia has operated in the modern era almost on the fringe of the industry, on the "left overs" as it were. Although our presence in the Antarctic waters may not have altered the fate of the humpback, and may in fact have hastened its demise, we would at least have had a share of the Antarctic cake while it was available.

What, in retrospect, seem to have been the "great days" of whaling, which may not have been romantic to the protagonists but were at least an almost elemental battle between one force and another, have long since gone. Nowadays, whaling is as much an assembly-line process as the manufacture of cars, despite the dangers and discomforts which still exist. The pity is that another of the world's great natural resources is rapidly being destroyed for the profit of a comparative few, and that distrust and greed have prevented the whaling nations from making any reasonable attempt to husband these resources. Whaling is one instance, at least, in which mankind could have "farmed the sea."

Instead, "the companions have made a banquet of him, and have parted him among the merchants."

BIBLIOGRAPHY

Adelaide *Advertiser*, 1 October 1932.

Adelaide *Advertiser*, 17 June 1933.

Adelaide *Advertiser*, 22 March 1935.

Adelaide *Advertiser*, 13 October 1936.

Adelaide *Advertiser*, 8 December 1938.

Melbourne *Argus*, 28 April 1938.

Ancher, E. A., *Mosman's Bay. Romance of an Old Whaling Station.* (Royal Australian Historical Society: *Journal and Proceedings*, Vol. ii, Pt. 10, 1909).

Annual Report, Chief Inspector of Fisheries, Western Australia, 9 January 1914.

Annual Report, Western Australian Whaling Company, 1914.

Australian Historical Society *Journal and Proceedings*, Vol. 1 & 2, 1901-1909.

Australian Quarterly, December 1961.

Australian Whaling Commission, 7th Annual Report.

Beale, T., *The Natural History of the Sperm Whale.* London 1839.

Borrow, K. T., *Whaling at Encounter Bay* (Adelaide Pioneers Association of S.A.).

Brady, E. J., *Whaling at Twofold Bay* (*Australia Today*, 1910).

Brierly, Oswald (Diary, Mitchell Library).

Colonist, The (Newspaper) Hobart Town, 24 August 1832.

Crowther, Dr W. L., *Notes on Tasman Whaling* (Proceedings of the Royal Society of Tasmania, 1919).

Dunabin, T., Whalers, *Sealers and Buccaneers* (Royal Australian Historical Society *Journal and Proceedings*, Vol. XI, 1925).

Fisheries Newsletter, June, 1946.
Fisheries Newsletter, August 1948.
Fisheries Newsletter, March 1956.
Fisheries Newsletter, February, 1963 (Article D. J. Gates).
Fisheries Newsletter, July, 1967.
Gazette, Sydney, 15 August 1828.
Godwin's *Emigrant's Guide to Van Dieman's Land* (1823).
Historical Records of New South Wales, 1803, 1805.
Hodge, Chas. R., *Encounter Bay, The Miniature Naples of Australia*, 1932.
Japan Whaling Industry (Japan Whaling Association 1954).
Leigh, W. H., *Voyages and Travels*, 1840.
McRae, Georgiana, *Georgiana McRae's Journal*, 1841-1865.
Melville, Herman, *Moby Dick*.
Melville, Henry, *Van Dieman's Land* (1833).
Moore, H. P., *Notes on Early Settlers in South Aust. prior to 1836*. Proceedings Royal Geographical Society of Australasia, S.A. Branch XXV, 1923-4.
Newland, Simpson. *Paving the Way*, 1893.
O'Hara, James. *The History of New South Wales*, 1818.
Parker, H. W. *Rise and Progress of Van Dieman's Land*, 1832.
Parliamentary Debates, Western Australia, Vol. 79 (New series), 1928.
Register (Newspaper), 7 October 1872.
Robinson, C. *New South Wales, the Oldest and Richest of Australian Colonies*, 1873.
Sidney, Samuel. *The Three Colonies of Australia*, 1853.
Statistical Society of South Australia (*Report*, 1841).
Stephens, John. *The Land of Promise, being the Authentic and Impartial History of South Australia*, 1839.
Sydney Morning Herald (Newspaper), 12 September 1936.
Sydney Record, 24 February 1944.
Votes and Proceedings of the Parliament, Western Australia, Vol. II, 3 December-4 March 1915.
Wellings, H. P. *The Brothers Imlay* (Royal Australian Historical Society; Journals and Proceedings, 1931).
Wellings, H. P. *Benjamin Boyd in Australia, 1842-49*.
Wellings, H. P. *Shore Whaling at Twofold Bay*.
Wells, W. H. *Gazetteer of the Australian Colonies*, 1848.
Western Mail, 29 June 1912.
Western Mail, 13 July 1912.

PERIODICAL ARTICLES

Arranged chronologically after title of serial.

Royal Geographical Society of Australasia, South Australian Branch. *Proceedings*.
Vol. 5, p. 66, 1901-2.
Vol. 10, p. 158, 1907-8.

Vol. 22, pp. 15-40, *Description and history of the whaling industry*, 1837-
1872.   1920-21.
Vol. 32, pp. 113-123, *Encounter Bay, Whaling Days* by Charles R.
Hodge.   1930-31.
Vol. 22, at end—Facsimile of Old Whaling Act in South Australia,
2 March 1840.   1920-21.
South Australian Magazine.
Vol. 1 (4) pp. 113-117, October, 1841. *Encounter Bay, its anchorages,
fisheries and general capabilities.*
South Australian Register.
12 August 1837, p. 3, col. 2. *Dispute between Captain Blenkinsopp and
the South Australian Company.*
11 November 1837, 2nd ed., p. 1, col. 2. *Operations at Port Lincoln
and Encounter Bay.*
20 January 1838, p. 2, col. 1. *Establishments at Encounter Bay.*
22 September 1838, p. 3, col. 4. *Progress of operations at Thistle Island
and Encounter Bay.*
1 January 1842, p. 3. *A history of the whaling industry, 1837-1841.*
7 October 1872, p. 15, col. 5. *Operations at Ranford's fishery at Encounter
Bay.*
6 September 1879, p. 5, col. 5. *Reminiscences of the whaling industry at
Encounter Bay.*
Southern Australian.
7 August 1839, p. 3, col. 5. *French ship engaged at Encounter Bay.*

SOURCES

Australian Public Affairs Information Services; a subject index to current
literature.   1945-
Catalogues:
Barr Smith Library, University of Adelaide.
State Library of South Australia.   Archives.
Crowley, F. K.
South Australian History; a survey for research students.
Libraries Board of S.A., Adelaide, 1966.
Ferguson, J. A.
Bibliography of Australia. Vol. 2, 1831-Vol. 5, 1901.
Sydney and London, Angus and Robertson, 1941.
Gill, T.
Bibliography of South Australia, Adelaide, 1886.
Historical Abstracts, 1955-
Public Library of New South Wales—Mitchell Library. *Index to periodicals.*
Sydney, 1944-
Royal Geographical Society of Australasia. South Australian Branch.
Index to the proceedings, Vols. 1-40, 1885-6 to 1938-9. Adelaide, 1944.
State Library of South Australia. Archives. Research note 166.

INDEX